Textile processing with enzymes

Textile processing with enzymes

Edited by

A. Cavaco-Paulo and G. M. Gübitz

The Textile Institute

CRC Press
Boca Raton Boston New York Washington, DC

WOODHEAD PUBLISHING LIMITED
Cambridge England

Published by Woodhead Publishing Limited in association with The Textile Institute
Woodhead Publishing Ltd
Abington Hall, Abington
Cambridge CB1 6AH, England
www.woodhead-publishing.com

Published in North America by CRC Press LLC
2000 Corporate Blvd, NW
Boca Raton FL 33431, USA

First published 2003, Woodhead Publishing Ltd and CRC Press LLC

British Library Cataloguing in Publication Data
A catalogue record for this book is available from the British Library.

Library of Congress Cataloging in Publication Data
A catalog record for this book is available from the Library of Congress.

Woodhead Publishing ISBN 1 85573 610 1
CRC Press ISBN 0-8493-1776-2
CRC Press order number: WP1776

Typeset by SNP Best-set Typesetter Ltd., Hong Kong
Printed by TJ International, Cornwall, England

Contents

Preface

A. CAVACO-PAULO
University of Minho, Portugal
and
G. GÜBITZ
Graz University of Technology, Austria

The first use of enzymes in textile processing was reported in 1857 when starch-sized cloth was soaked with liquor containing barley. Later, in 1900, this process was slightly improved using malt extract, but only in 1912 with the use of animal and bacterial amylases was the process of enzymatic desizing introduced into many textile factories. Interestingly, amylases remained the only enzymes applied in textile wet processing for almost 70 years. In the late 1980s, cellulases were introduced with great success for depilling and defuzzing cellulose-based fabrics as well as to age garments made from materials like denim to obtain the stone-washed look. Since the early 1990s, catalases have been introduced to destroy hydrogen peroxide after bleaching, reducing the consumption of water. Pectin degrading enzyme products have been commercialised for cotton processing to replace traditional alkaline scouring. Intense investigations are being conducted on new enzyme applications for almost all cotton processing steps and for modification of cellulosic, proteic and synthetic fibres. Textile processing with enzymes is therefore a new emerging and multidisciplinary area. Engineers with knowledge and basic understanding in both textile technology and enzymology will help to introduce these environmentally-friendly processes more extensively to the industry. However, only little information about enzymes for textile processing can be found in educational programmes or in the literature.

This book was put together to generate a basic understanding of enzymes, textile materials and process engineering. It can serve as a textbook for everyone interested in the subject; students, scientists and engineers alike with a basic background in either textiles, biotechnology, chemistry or engineering. The book covers all relevant aspects of textile processing with enzymes, from the chemical constitution and properties of textile materials as potential substrates for enzymes, to processing of these materials, and from basic biochemistry and enzymology to industrial application of these biocatalysts.

Chapter 1 deals with the fundamental aspects of enzymes determining catalytic properties. It is intended to provide a basis for the understanding of many aspects related to the application of enzymes considered in subsequent chapters. Chapter 2 gives an overview of non-fibrous and fibrous materials as substrates for enzymes. Included is a discussion on dyes, sizes, textile fibres and textile auxiliaries that might influence enzymatic reactions. Chapter 3, about catalysis and processing, gives an overview about the function and application of enzymes used in textile processing. Basic thermodynamics and enzyme kinetics, function of textile-processing enzymes, homogenous and heterogeneous catalysis and important applications of enzymes in textile wet processing are addressed. Chapter 4 gives insights into process engineering and describes major problems in the industrial applications of enzymes in textiles. Important facts about the influence of mass transfer are described. Chapter 5 discusses practical aspects of handling enzymes, like enzyme activity. Operational and storage stabilities are discussed in detail as well as health and safety issues. The last chapter, Chapter 6, deals with effluent treatment and the potential use of enzymes therein.

Contributors

(* indicates main point of contact)

Editors (Preface and Chapter 3)

Prof. A. Cavaco-Paulo*
University of Minho
4800 Guimarães
Portugal

Tel: +351 253 510271
Fax: +351 253 510293
E-mail: artur@det.uminho.pt

Prof. G. Gübitz
Graz University of Technology
Petersgasse 12
A-8010 Graz
Austria

Tel: +43 316 873 8312
Fax: +43 316 873 8815
E-mail: guebitz@ima.tu-graz.ac.at

Chapter 1
Dr R. O. Jenkins
School of Molecular Sciences
De Montfort University
The Gateway
Leicester
LE1 9BH

Tel: +44 (0) 116 250 6306
Fax: +44 (0) 116 257 7235
E-mail: roj@dmu.ac.uk

Chapter 2
Dr G. Buschle-Diller
Auburn University
Textile Engineering Department
115, Textile Building
Auburn
Alabama 36849-5327
USA

Tel: +1 334 844 5468
Fax: +1 334 844 4068
E-mail: giselabd@Eng.auburn.edu

Chapter 4
Dr ir. V. A. Nierstrasz* and
Prof. dr. ir. M. M. C. G. Warmoeskerken
Textile Technology Group
Department of Science and Technology
University of Twente
PO Box 217
NL-7500 AE Enschede
The Netherlands

Tel: +31 (0) 53 489 2899
Fax: +31 (0) 53 489 3849
E-mail: v.a.nierstrasz@utwente.nl

Chapter 5
Dr H. B. M. Lenting
TNO Institute for Industrial Technology

Centre for Textile Research
PO Box 337
7500 AH Enschede
The Netherlands

Tel: +31 53 486 0490
Fax: +31 53 486 0487
E-mail: h.lenting@ind.tno.nl

Chapter 6

Dr J. Binkley*
UMIST
P O Box 88
Manchester
M60 1QD
UK

Tel: +44 (0) 1204 598873
E-mail: jb450@tiscali.co.uk

A. Kandelbauer
Graz University of Technology
Petersgasse 12
A-8010 Graz
Austria

Tel: +43 316 873 8312
Fax: +43 316 873 8815

1
Enzymes

RICHARD O. JENKINS
De Montfort University, UK

1.1 Introduction

Enzymes are biological catalysts that mediate virtually all of the biochemical reactions that constitute metabolism in living systems. They accelerate the rate of chemical reaction without themselves undergoing any permanent chemical change, i.e. they are true catalysts. The term 'enzyme' was first used by Kühne in 1878, even though Berzelius had published a theory of chemical catalysis some 40 years before this date, and comes from the Greek *enzumé* meaning 'in (*en*) yeast (*zumé*)'. In 1897, Eduard Büchner reported extraction of functional enzymes from cells. He showed that a cell-free yeast extract could produce ethanol from glucose, a biochemical pathway now known to involve 11 enzyme-catalysed steps. It was not until 1926, however, that the first enzyme (urease from Jack-bean) was purified and crystallised by James Sumner of Cornell University, who was awarded the 1947 Nobel Prize. The prize was shared with John Northrop and Wendell Stanley of the Rockefeller Institute for Medical Research, who had devised a complex precipitation procedure for isolating pepsin. The procedure of Northrop and Stanley has been used to crystallise several enzymes. Subsequent work on purified enzymes, by many researchers, has provided an understanding of the structure and properties of enzymes.

All known enzymes are proteins. They therefore consist of one or more polypeptide chains and display properties that are typical of proteins. As considered later in this chapter, the influence of many chemical and physical parameters (such as salt concentration, temperature and pH) on the rate of enzyme catalysis can be explained by their influence on protein structure. Some enzymes require small non-protein molecules, known as cofactors, in order to function as catalysts.

Enzymes differ from chemical catalysts in several important ways:

1. Enzyme-catalysed reactions are at least several orders of magnitude faster than chemically-catalysed reactions. When compared to the

corresponding uncatalysed reactions, enzymes typically enhance the rates by 10^6 to 10^{13} times.
2. Enzymes have far greater reaction specificity than chemically-catalysed reactions and they rarely form byproducts.
3. Enzymes catalyse reactions under comparatively mild reaction conditions, such as temperatures below 100°C, atmospheric pressure and pH around neutral. Conversely, high temperatures and pressures and extremes of pH are often necessary in chemical catalysis.

1.1.1 In this chapter

This chapter is concerned mainy with the fundamental aspects of enzymes that determine their properties and catalytic capabilities. It is intended to provide a sound basis for understanding of many of the applied aspects of enzymes considered in subsequent chapters in this text. Given the wealth of fundamental knowledge on enzymes, it is only possible here to provide a perspective on each of the topics. Some of the topics will be considered in more detail, or from a different perspective, later on in the text.

Section 1.2 deals with the classification and nomenclature of enzymes. It considers some of the rules that form the basis of a rational system classification and naming enzymes, and provides examples of enzymes in each of the six main classes. Much of the chapter is devoted to protein structure (Section 1.3) because this ultimately defines the properties of enzymes, such as substrate specificity, stability, catalysis and response to physical and chemical factors. Protein structure is considered at all levels of organisation, from the 'building blocks' (amino acids) of proteins, through backbone conformations and three-dimensional shapes, to enzymes having more than one sub-unit. Consideration of the forces that stabilise protein molecules follows (Section 1.4) and the strengths of the various bonds are compared in relation to level of protein structure. Section 1.5 briefly describes some of the basic properties of proteins, such as chemical reactions with reactive amino acid groups, the acid–base properties of enzymes and some other factors (temperature and pH) that influence protein solubility and catalytic activity. Cellular biosynthesis of proteins is described in Section 1.6, with the emphasis very much on the process of reading the genetic code to synthesising a chain of amino acids in the correct predetermined sequence. This is followed by a section (1.7) on enzymatic modification of proteins within cells *after* they have been synthesised. Such post-translational modification influences the structural stability or activity of enzymes. Section 1.8 considers enzymatic catalysis, with the emphasis on enzyme substrate specificity and the requirement of some enzymes for the presence of non-proteinaceous compounds for catalytic activity. Comments on future trends (Section 1.9) and recommendations for further reading (Section 1.10) are also included. Papers from the primary literature have not been referred to; rather, a list of relevant books and review articles are provided at the end of the chapter.

1.2 Classification and nomenclature of enzymes

Organisms – whether animal, plant or microorganism – are both complex and diverse. In biological systems, thousands of different types of reactions are known to be catalysed by different enzymes; many more are yet to be discovered. The diversity of enzymes is, therefore, enormous in terms of type of reaction(s) they catalyse, and also in terms of structure. Enzymes range from individual proteins with a relative molecular mass (RMM) of around 13 000 catalysing a single reaction, to multi-enzyme complexes of RMM several million catalysing several distinct reactions.

Given such diversity, it is essential to have a rational basis for classification and naming of enzymes. Currently, it is the Nomenclature Committee of the International Union of Biochemistry and Molecular Biology (NC-IUBMB) that considers these matters and gives recommendations to the international scientific community.

1.2.1 General rules

Enzymes are principally classified and named according to the **chemical reaction** they catalyse, as this is the specific property that distinguishes one enzyme from another. It is the observed chemical change produced by the complete enzyme reaction that is used for this purpose, i.e. the overall reaction, rather than the formation of intermediate complexes of the reactants with the enzyme. Some notable consequences of this system are:

- A systematic name cannot be given to an enzyme until the chemical reaction is known. This applies, for example, to enzymes that catalyse an isotopic change to a molecule that indicates one step in the overall reaction, but the reaction as a whole remains unknown.
- An enzyme name is assigned not to a single enzyme protein but to a group of proteins with the same catalytic property. Some exceptions exist, where more than one name is assigned to enzymes with the same catalytic property because the reaction is so different in terms of substrate specificity or mechanism. Other exceptions include acid and alkaline phosphatases. These enzymes carry out the same reaction but at widely different pH values.
- Enzymes from different sources – such as animal, plant and microorganisms – are classified as one entry.

- To classify an enzyme it is occasionally necessary to choose between alternative ways of regarding the chemical reaction. In general, the alternative selected should reduce the number of exceptions.
- The direction of the chemical reaction needs to be considered, since all reactions catalysed by enzymes are reversible. For simplicity, the direction chosen should be the same for all enzymes in a given class even if this direction has not been shown for all of the enzymes.

The Enzyme Commission of the International Union of Biochemistry, in its report of 1961, devised a rational system for classification of enzymes and assigning code numbers to them based on the reaction catalysed. The code numbers, prefixed by EC, are now used widely and contain four elements separated by points:

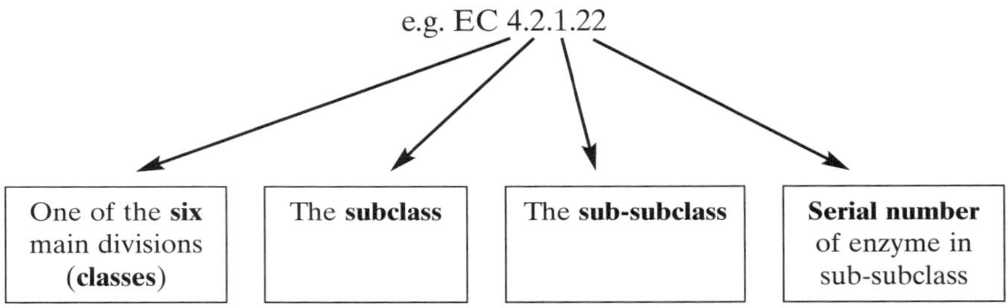

There are six classes of enzymes that are distinguished by the first digit of the EC code (Table 1.1). The second and third digits describe further the type of reaction catalysed. These digits are defined for each of the separate main classes of enzymes and there is no general rule that applies to their meaning. Enzymes that catalyse very similar reactions, e.g. enzymes that cleave C—O bonds in a substrate molecule, will have the same first three digits in their EC code. They will, however, have different fourth digits that define the actual substrate for the reaction.

A consequence of enzymes being classified according to the chemical reaction they catalyse is that **isoenzymes** (different enzymes catalysing identical reactions) carry the same four digit EC classification number. There are, for example, five different isoenzymes of lactate dehydrogenase in the human body and the EC code does not provide a means of distinguishing between them. Rather, the particular isoenzyme and its source (e.g. mammalian heart) have also to be specified.

1.2.2 Recommended and systematic names

The Enzyme Commission has recommended that there should be 'systematic' as well as 'trivial' (working) nomenclatures for enzymes; examples for

Table 1.1 Classification and nomenclature for the six classes of enzymes

1. **Oxidoreductases**: enzymes that catalyse oxidoreductase reactions
   2nd EC digit: indicates group in the hydrogen donor (substrate oxidised), e.g. —CHOH—, aldehyde, keto
   3rd EC digit: indicates type of acceptor involved, e.g. a cytochrome, molecular oxygen, an iron–sulphur protein, etc
   Systematic name: *donor:acceptor oxidoreductase*
   Recommended name: *donor:dehydrogenase* (*reductase* as alternative; *oxidase* where O_2 is acceptor)
   e.g. alcohol dehydrogenase (trivial); alcohol NAD$^+$ oxidoreductase (EC 1.1.1.1)
2. **Transferases**: enzymes transferring a group
   2nd EC digit: indicates group transferred, e.g. methyl, glycosyl, phosphate
   3rd EC digit: further information on group transferred, e.g. hydroxymethyl
   Systematic name: *donor:acceptor grouptransferase*
   Recommended name: *acceptor grouptransferase*
   e.g. glucokinase; ATP glucose phosphotransferase (EC 2.7.1.2)
3. **Hydrolases**: enzymes that catalyse cleavage of C—O, C—N, C—C and some other bonds
   2nd EC digit: indicates nature of bond hydrolysed, e.g. ester, glycosyl
   3rd EC digit: indicates nature of substrate, e.g. carboxylic ester, thiolester
   Systematic name: *substrate:hydrolase*
   Recommended name: *substrate* with suffix *-ase*
   e.g. carboxypeptidase A (EC 3.4.17.1)
4. **Lyases**: enzymes that cleave C—C, C—O, C—N and other bonds by elimination, leaving double bonds or rings, or add groups to double bonds
   2nd EC digit: indicates the bond broken
   3rd EC digit: further information on group eliminated, e.g. CO_2, H_2O
   Systematic name: *substrate group-lyase* (hyphen included)
   Recommended names: e.g. *decarboxylase, dehydratase* (in case of elimination of CO_2 and H_2O); *synthase* used if reverse reaction described
   e.g. pyruvate decarboxylase; pyruvate-lyase (EC 4.1.1.1)
5. **Isomerases**: enzymes that catalyse geometric or structural changes within one molecule
   2nd EC digit: indicates type of isomerism, e.g. racemase, epimerase, *cis-trans* isomerase
   3rd EC digit: indicates type of substrate
   Systematic name: *substrate:type of isomerism*
   Recommended name: *substrate:isomerase*
   e.g. maleate isomerase; maleate *cis-trans* isomerase (EC 5.2.1.10)
6. **Ligases**: enzymes catalysing the joining of two molecules coupled with hydrolysis of a diphosphate bond in ATP (or similar triphosphate)
   2nd EC digit: indicates the bond formed, e.g. C—O, C—S, C—N
   3rd EC digit: (only used in the C—N ligases)
   Systematic name: *X:Y ligase (ADP-forming)*
   Recommended name: *X:Y ligase* (previously *synthetase* was used)
   e.g. pyruvate carboxylase (trivial); pyruvate carboxyligase (ADP forming) (EC 6.4.1.1)

each of the six classes of enzymes are given in Table 1.1. The systematic name describes the action of an enzyme as exactly as possible, whereas the trivial name is sufficiently short for general use and is often a name already in common use. The Enzyme Commission-recommended trivial names for new enzymes are often condensed versions of systematic names.

Since enzymes are divided into groups according to the type of reaction catalysed, this and the name(s) of the substrate(s) are the basis for systematic naming of individual enzymes. It is also the basis for classification and code numbers. Names of enzymes, especially those ending in *ase*, generally refer to single enzymes and are not applied to systems containing one or more enzymes. When an overall reaction involving more than one enzyme is named, the word 'system' is included in the name. For example, the 'succinate oxidase system' is used to describe the enzymatic oxidation of succinate involving succinate dehydrogenase, cytochrome oxidase and several intermediate carriers.

General rules for systematic names and guidelines for recommended names, as well as rules and guidelines for particular classes of enzymes, are available at Enzyme Nomenclature Database at the Swiss Institute of Bioinformatics (http://www.espasy.ch/enzyme).

1.3 Protein structure

1.3.1 Overview

Proteins consist of one or more polypeptides and each polypeptide is a chain of amino acids linked together by peptide bonds. A different gene codes for each polypeptide and determines the sequence of amino acids of the polypeptide. Polypeptide chains fold up when synthesised to form a unique three-dimensional shape (conformation), determined by their amino acid sequences. Multiple weak interactions stabilise the conformation of polypeptides and factors (such as pH, heat and chemicals) that disrupt these interactions distort the polypeptide's conformation. Enzymes lose their functional activity when their three-dimensional conformation is distorted in this manner, through enzyme denaturation. This demonstrates a clear dependence of enzyme functioning upon protein structure.

There are two main types of proteins: 'fibrous' and 'globular'. **Fibrous** proteins normally have a structural role in biological systems. They are insoluble in water and are physically durable/strong. The three-dimensional structure of fibrous proteins is relatively simple and usually elongated. Examples of fibrous proteins are:

- α-Keratin: the main protein of hair, nails, wool, horn and feathers
- β-Keratin: the main structural component of silk and spider's web

- Collagen: a major protein of cartilage, tendons, skin and bones
- Elastin: a protein found in ligaments in the walls of arteries.

Globular proteins are generally soluble in water and can often be crystallised from solution. They have a more complex three-dimensional structure and tend to adopt an approximate spherical shape in which the amino acid chain is tightly folded. Globular proteins have functional roles in biological systems and all enzymes are globular proteins. Proteins are also categorised as 'simple' or 'conjugated'. **Simple** proteins are composed entirely of amino acids, while **conjugated** proteins contain one or more other materials bound to one or more of the amino acid residues. Examples of conjugated proteins and their bound components are:

- Glycoproteins – carbohydrate
- Metalloproteins – metal ions
- Lipoproteins – lipids
- Nucleoproteins – nucleic acids
- Flavoproteins – flavin nucleotides

1.3.2 Amino acids – the 'building blocks' of proteins

Amino acids are organic molecules that contain an amino group (primary $—NH_2$; secondary >NH) and a carboxyl group ($\mathrm{O{=}\overset{|}{C}{—}OH}$ or —COOH). There are 20 commonly occurring amino acids. All except one has a central (α) carbon atom, to which is attached a primary amino group ($—NH_2$), a carboxyl group (—COOH), a hydrogen atom and a side group or chain (R); the side groups are different in all amino acids. Proline is unique because it lacks a primary amino group; instead it contains a secondary amino group (>NH). In proline the side group is curled round so that the nitrogen and the α-carbon atoms form part of a non-polar and fully saturated five-membered imino ring; proline is termed an imino acid. Representations of the generalised structure of amino acids are shown in Fig. 1.1.

Side chains (R-groups) of α-amino acids are polar or non-polar. The structure of the polar molecules may be stabilised by hydrogen bonding in aqueous solution; they display ionic character and are therefore hydrophilic and soluble in water. Conversely, non-polar molecules are relatively insoluble in water, but more soluble in organic solvents. The categorisation of amino acids according to the hydrophobic or hydrophilic character of the side chains is shown in Fig. 1.2. Phenylalanine, tryptophan and tyrosine are termed aromatic amino because the R-group has a six-membered aromatic benzene ring, whereas hystidine has a five-membered imidazole ring. The double-ringed R-group of tryptophan is called indole and in tyrosine the ring is linked to —OH to form a phenolic group. As mentioned earlier,

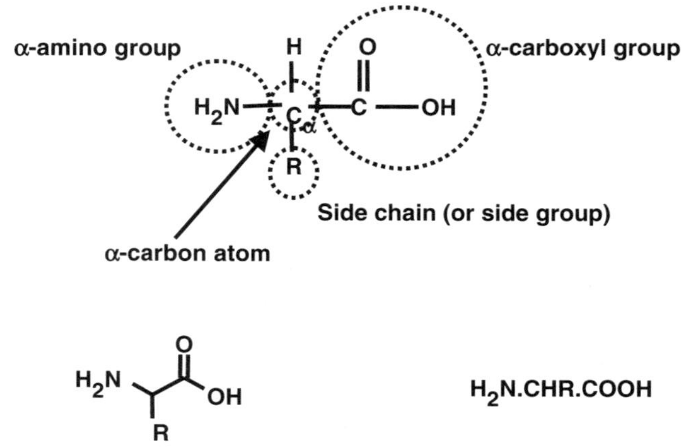

1.1 Representations of the generalised structure of amino acids.

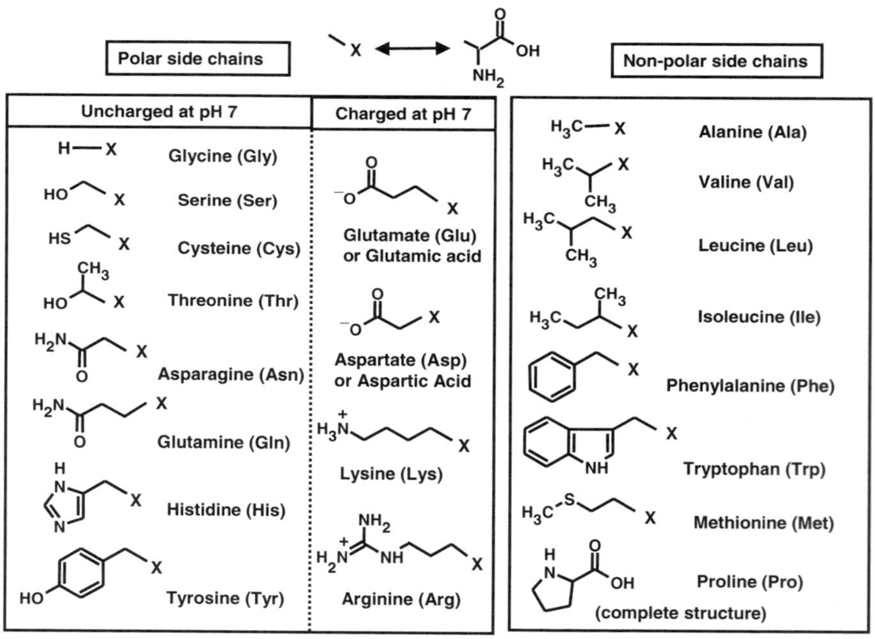

1.2 Categorisation of amino acids according to hydrophobic or hydrophilic character of side chains.

proline – the only other amino acid containing a ring structure – is an imino acid. Several of the hydrophobic side chains are branched-chain aliphatic hydrocarbons. Glutamic acid and aspartic acid have hydrophilic side chains containing carboxyl groups, which are converted to amide groups in

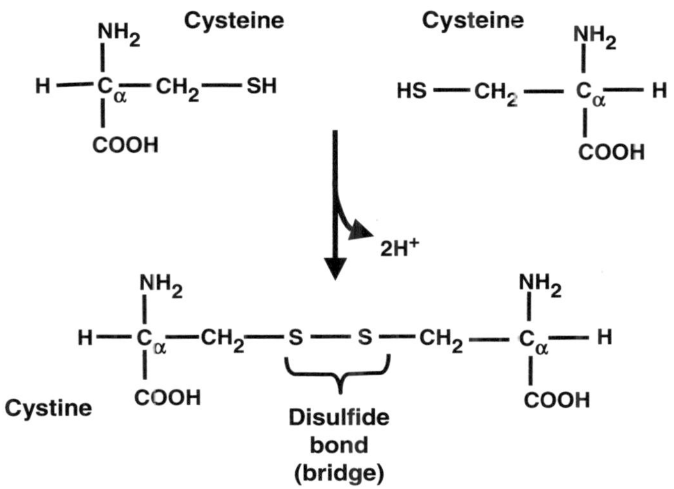

1.3 Oxidation of the sulfydryl groups of cysteine to form cystine.

asparagine and glutamine, respectively. Arginine has a guanidine side chain that, in common with the side chain of lysine, contains an amino group. Cysteine and methionine are sulfur-containing amino acids. In cysteine the sulfydryl group (—SH) oxidises readily to form the dimeric compound cystine, which comprises two cysteine residues linked by a disulfide bridge or bond (Fig. 1.3).

With the exception of glycine, all of the common amino acids exist as optical isomers. These are two mirror image forms of an amino acid that cannot be superimosed by rotation of the molecule. They arise because amino acids (with the exception of glycine) contain at least one α-carbon atom covalently linked to four different atoms or groups. The molecule is therefore asymmetric because no plane drawn through the α-carbon atom can divide the molecule into two parts that are exact mirror images. It follows that two mirror image forms of the complete molecule can exist. The isomers are termed optical isomers because one will rotate the plane of polarized light to the left, and the other to the right. They are termed D- and L-isomers for the sake of distinction, although this does not indicate how they affect the plane of polarised light. Figure 1.4 illustrates the arrangement of groups around an asymmetric α-carbon of an amino acid. Proteins are almost exclusively composed of L-amino acids. The reason for this is that protein biosynthesis is mediated by enzymes that distinguish the optical isomers of amino acids in a solution containing both L- and D-forms (enantiomers); steriospecificity of enzyme action is considered in more detail in Section 1.8.1.

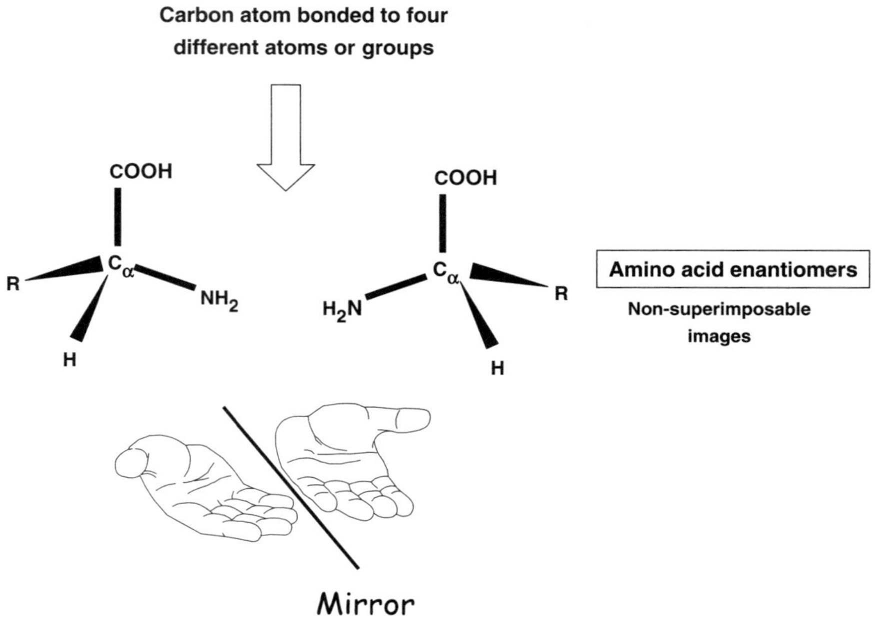

1.4 Arrangement of groups around an asymmetric α-carbon of an amino acid.

1.3.3 Primary structure of proteins

Amino acids (monomers) are joined together by peptide bonds to give proteins. Addition of increasing numbers of amino acids gives peptides and then polypeptides. If the RMM of the chain is more than 5000, the molecule is usually referred to as a polypeptide rather than a peptide. The primary structure of a polypeptide refers to the amino acid sequence, together with the positioning of any disulfide bonds that may be present.

The peptide (amide) bond that joins amino acids together to form a polypeptide is formed by elimination of water, i.e. a condensation reaction (Fig. 1.5). The polypeptide chain formed (the residue) has one free carboxy and one free amino group at opposite ends of the molecule, known as the carboxy and amino termini, respectively. Peptide bonds are rigid, being stabilised by resonance, i.e. the amide nitrogen lone pair of electrons is delocalised across the peptide linkage. The bond can be thought of as having an intermediate form between the two extremes (*cis* and *trans* forms). However, in most instances steric interference between the amino acid side groups and the α-carbon atoms of adjacent amino acid residues means that the *trans* form (R-group lies on opposite sites of the polypeptide chain; Fig. 1.5) is around 1000-fold more common than the *cis* form. This minimizes

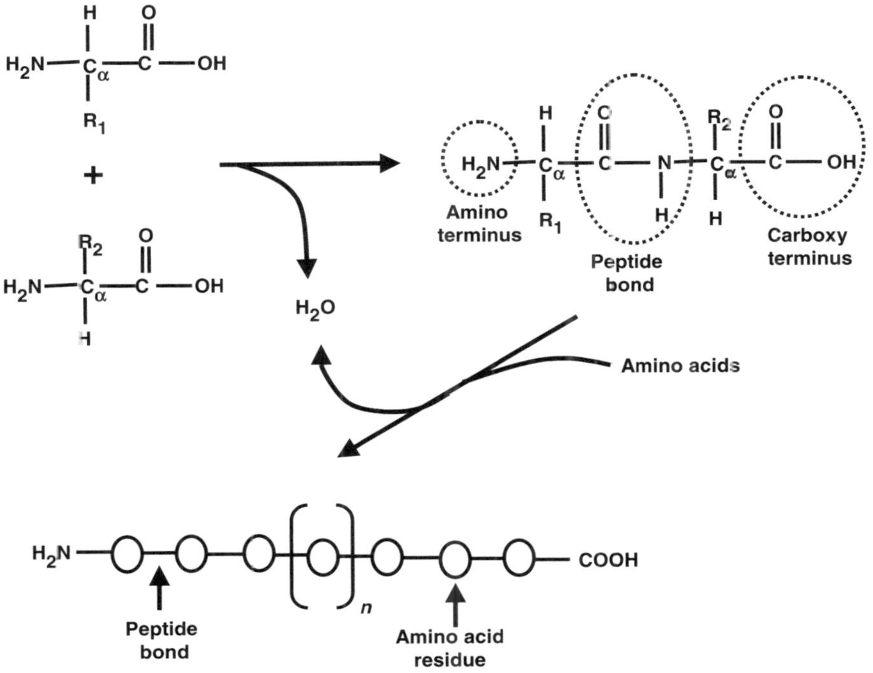

1.5 Linking of amino acids by peptide bond formation.

steric interference between the R groups in the peptide chain. In the case of proline, however, the unusual side group of this amino acid allows the peptide bond to adopt the *cis* configuration.

While the peptide bond has a rigid and planar structure, the other bonds in the polypeptide are free to rotate (Fig. 1.6). Rotation about the N—C_α bond is denoted by Φ (phi) and the C_α—C bond by ψ (psi). When the amino acids are in the *trans* form the polypeptide chain is fully extended and, by convention, these rotation angles (also known as dihedral or torsion angles) are defined as being 180°. In principal, each bond can rotate in either direction and have values −180° and +180°. However, steric hindrance between the atoms of the polypeptide backbone and those of amino acid side chains restricts the degree of rotation and the majority of ψ and Φ combinations are excluded. Rotational freedom around glycine residues is relatively high since steric hindrance is minimised by the small R-group (i.e. hydrogen). Conversely, rotation around proline residues is restricted owing to the unusual structure of the side chain of this amino acid. The ψ and Φ angles, together with the fixed ω angles of all the residues, define the conformation of the main chain (backbone) of the polypeptide. Certain combinations of ψ and Φ form relatively stable regularly-shaped backbone

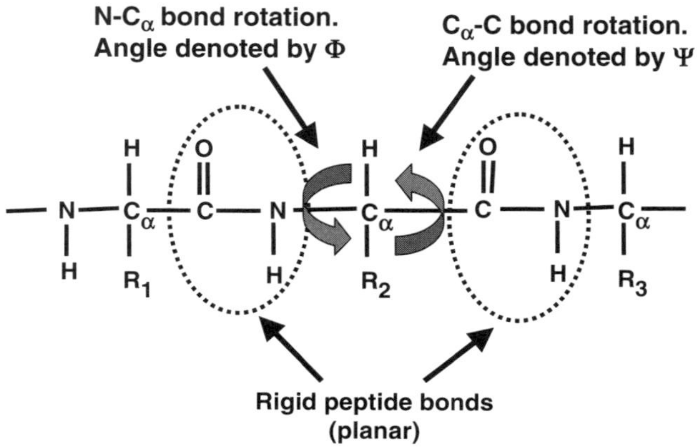

1.6 Rotation in a section of polypeptide chain.

conformations called **secondary structures**. These angles also have a major influence on the final three-dimensional shape (**tertiary structure**) of the polypeptide.

1.3.4 Secondary structure of proteins

Secondary structure is the local spatial conformation of the back bone of a polypeptide, excluding the side chains (R-groups) of the amino acids. Three secondary structures are common in proteins: α-helices, β-sheets and β-turns. These structures are commonly formed because they minimise steric interference between adjacent side-chain groups and maximise formation of intermolecular hydrogen bonds that are closely situated in the primary structure. These bonds stabilise the secondary structures. Parts of the polypeptide backbone that do not have recognisable secondary structures are referred to as 'random coils', or sometimes merely as 'coils', and tend to have more flexibility of movement in solution compared to the secondary structures.

α-Helices are right handed helices containing 3.6 amino acid residues in a full turn. This arrangement is stabilised by hydrogen bonding between carboxyl and amino groups in the polypeptide backbone; every —C=O group forms a hydrogen bond with the —N—H group of the amino acid residue four positions ahead of it in the helix. This means that the hydrogen bonds link together a 13-atom length of polypeptide backbone and thus the α-helice is described as 3.6_{13}. α-Helices are found in both fibrous and globular proteins. In enzymes (globular proteins), the average length of a helical region is three turns, but may vary from a single turn to more than ten consecutive turns. Sometimes a single turn of 3_{10} is found at the end of α-helices.

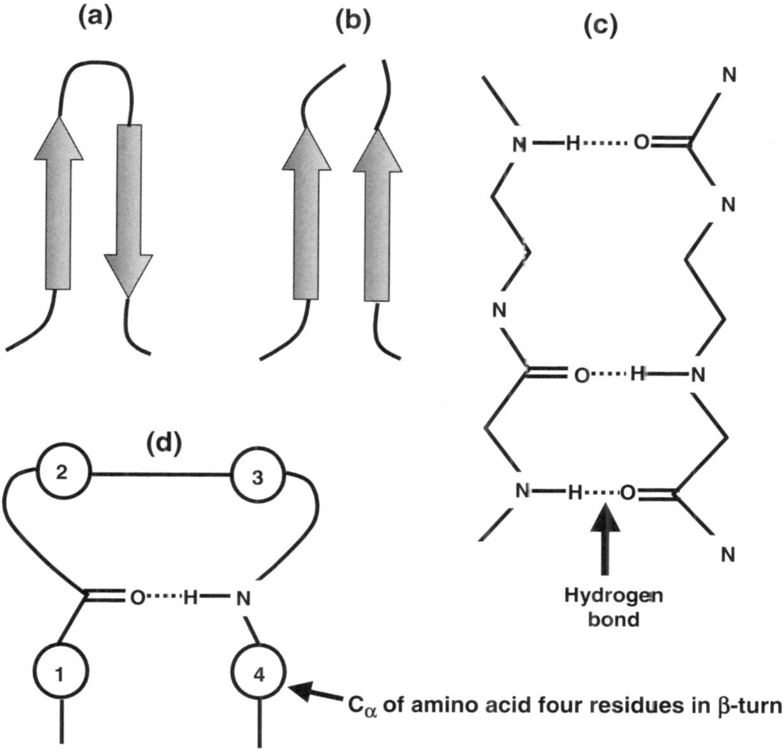

1.7 β-Sheets made up of two β-strands of the polypeptide chain. Orientation of strands in (a) an anti-parallel and (b) a parallel β-sheet. (c) Hydrogen bonding stabilising an anti-parallel β-sheet. (d) Hydrogen bonding in a β-turn over four amino acid residues.

The side groups of the amino acid residues project outwards, thus the helix has a hollow core.

β-Sheets are relatively extended sections of the polypeptide backbone. They are made up of two β-strands of the polypeptide chain. Each strand is usually five to ten amino acid residues in length, with adjacent peptide groups tilted in alternate directions, giving a zig-zag or pleated conformation; the two strand sections are also termed **β-pleated sheets**. The >N—H and >C=O groups of the amino acid residues that point out at approximate right angles to the extended polypeptide backbone, form hydrogen bonds between two β-strand regions from the same polypeptide or from different polypeptides held in close proximity. Parallel β-sheets occur when the polypeptide chains run in the same direction (i.e. N to C), whereas anti-parallel β-sheets have chains running in opposite directions (Fig. 1.7a,b). The latter structure is the most stable of the two because hydrogen bonding is more effective in the anti-parallel conformation (Fig. 1.7c). A mixed β-sheet contains both parallel and anti-parallel strands.

β-Sheets, like α-helices, are found in both fibrous and globular proteins and are essentially linear structures. In order for polypeptides to fold to form a compact tertiary structure, changes in direction of the polypeptide backbone are necessary. Such changes in direction occur in 'loop' regions of the polypeptide between stretches of regular secondary structures (β-sheets and/or α-helices). Loops are generally found on the surface of the polypeptide since they are rich in charged (polar) amino acid residues that bond with surrounding water molecules. In globular proteins, which have a roughly spherical shape, the commonest type of loop structure is the **β-turn** (also known as β-bend or reverse turn). The β-turn introduces a 180° change in direction of the polypeptide chain over four amino acid residues (Fig. 1.7d). Glycine and proline are the prominent amino acids in β-turns and give rise to the change in direction of the polypeptide chain. Proline naturally introduces a twist into the polypeptide chain owing to its ring structure side chain. Glycine, owing to its small side group, minimises steric interference and occupies the restricted space available. A hydrogen bond forms between the —C=O of the first amino acid residue and the NH of the fourth residue, which helps to stabilise the loop (Fig. 1.7d).

1.3.5 Tertiary structure of proteins

The tertiary structure of a protein is the exact three-dimensional shape of the folded polypeptide, i.e. the positioning in space of all the atoms in the polypeptide relative to each other. Polypeptides comprising more than around 200 amino acid residues often have two or more structural subunits, known as 'domains'. These are tightly folded sub-regions of a single polypeptide, which are connected by more flexible and extended regions of the polypeptide. Domains are usually comprised of 'structural motifs' (sometimes referred to as 'supersecondary structures'), which are secondary structures occurring as closely associated structures. Figure 1.8 illustrate some common structural motifs. The three-dimensional shape of an actual enzyme, glycosyl hydrolase, is illustrated in Fig. 1.9.

Although polypeptides are most often characterised according to their biological activity or function, they can also be characterised based on the domain structure of the polypeptide. There are three main domain types:

1 α-domain. Comprised exclusively of α-helices, e.g. arranged as a four-helical bundle motif.
2 αβ-domain. Most common domain types in proteins. Comprised of parallel β-sheets surrounded by stretches of α-helix.
3 β-domain. Comprised of a core of anti-parallel β-sheets.

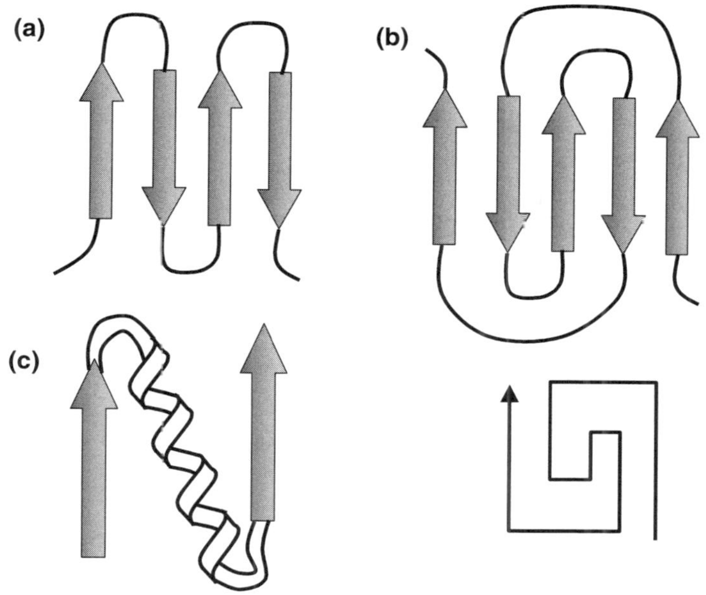

1.8 Some common structural motifs (supersecondary structures) of proteins: (a) β-meander; (b) Greek key (common desigr in Greek architecture); (c) βαβ unit.

The tertiary structure of a protein is usually so compact that there is virtually no room for water. For non-membrane associated proteins, non-polar side chains of the amino acid residues are nearly all located in the middle of the structure, whereas polar side chains are located on the surface.

X-ray diffraction, nuclear magnetic resonance (NMR) and electron microscopy are techniques used to elucidate the three-dimensional structures of proteins, and each have their advantages and limitations. Electron microscopy provides information only at low-resolution and its principal application has been to provide information on the overall three-dimensional shape of large proteins or protein aggregates. Both X-ray defraction and NMR, however, provide high-resolution information at the atomic level. X-ray diffraction involves bombarding a crystal of a protein with electromagnetic radiation of a wavelength approximately the same as the dimensions of the atoms in a protein, i.e. X-rays, 10^{-10} m. Some of the X-rays are diffracted by the atoms in the protein crystal, giving a diffraction pattern that reflects the three-dimensional structure of the protein molecules in the crystal. This is a powerful technique whose main limitation is the requirement for the protein to be in crystalline form. The vast majority of globular proteins do not crystallise readily; they are generally large, have

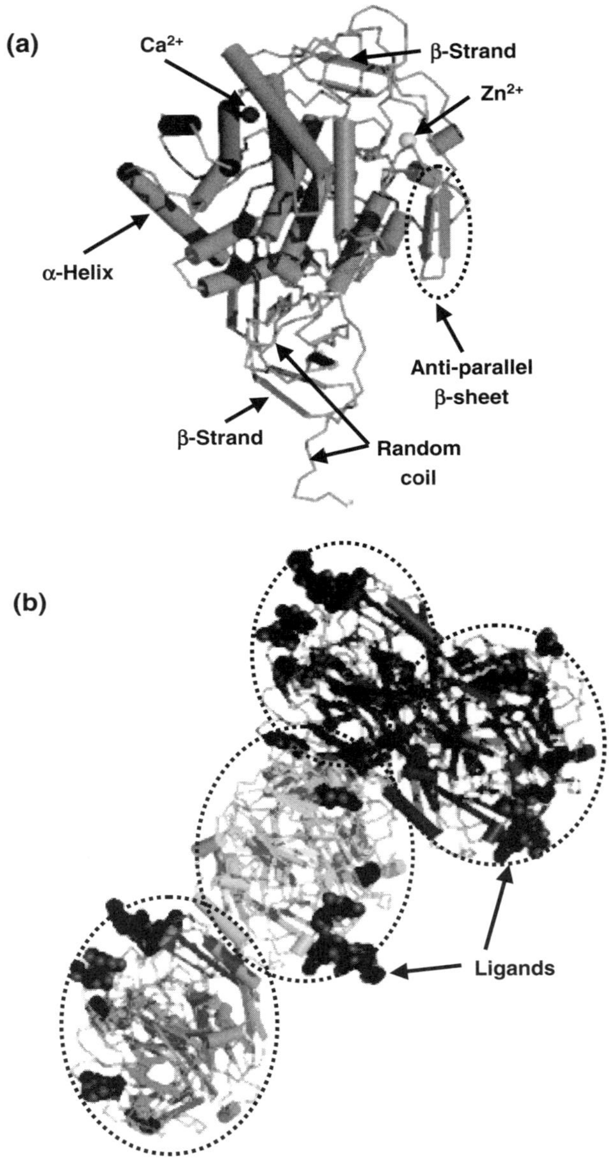

1.9 Three-dimensional structure of glycosyl hydrolase (synonym: cellulase endo-1,4-β-glucanase D) from *Clostridium thermocellum*. Main-polypeptide chain: 541 amino acid residues comprising 17α-helices, 51β-strands; associated metals, $3 \times Ca^{2+}$, $1 \times Zn^{2+}$. Data obtained from the Protein Data Bank.[20]

irregular surfaces and tend to hold significant amounts (>30%) of water within their structures.

NMR can be used to determine protein structure in solution. The technique involves subjecting a sample of the protein solution to a strong magnetic field. This causes the spin of the atomic nuclei in the protein (such as ^{13}C and ^{1}H) to align along the magnetic field and, if the appropriate radiofrequency energy is applied, the alignment can be converted to an excited state. When the nuclei then revert back to an unexcited state, radiofrequency radiation is emitted whose exact frequency is influenced by the molecular environment of the individual nuclei. Detection and measurement of the emitted radiation therefore provides information on the three-dimensional structure of proteins. The main limitation of NMR in this application is the complexity of the data that is generated and the technique is generally applied only to relatively small proteins, having a relative molecular mass of <25 000.

1.3.6 Quaternary structure of proteins

Some proteins are made up of more than one polypeptide chain, or subunit (Fig. 1.9). The quaternary structure of a protein describes subunit structure. Proteins that lack quaternary structure are referred to as **monomeric**, those having two-subunits as **dimeric**, three as **trimeric**, four as **tetrameric**, etc. Different types of subunits are assigned Greek letters, with the number of each subunit indicated by a subscripted number. For example, in the case of phosphorylase kinase, the subunit composition is described as $\alpha_4\beta_4\gamma_4\delta_4$, indicating that there are four each of four different subunits, i.e. 16 separate polypeptides.

The vast majority of enzymes are oligomeric enzymes, with molecular weight in excess of 35 000. Some oligomeric enzymes form **isoenzymes**, which are different enzymes catalysing the same chemical reaction; they carry the same four digit EC classification number (see Section 1.2). Isoenzymes typically have different kinetic properties and commonly arise through the occurrence of alternative subunits in an oligomeric enzyme. For example, mammalian lactate dehydrogenase is a tetramer, in which the subunits may be of two types: the heart-type (H) and the muscle-type (M). Five possible combinations of subunits (isoenzymes) exist: H_4, H_3M_1, H_2M_2, H_1M_3, M_4. The five isoenzymes have different kinetic characteristics. Different tissues have characteristic ratios of the five isoenzymes that suit particular metabolic needs.

1.3.7 Enzymes that lack a quaternary structure

Enzymes that lack a quaternary structure are known as **monomeric** enzymes, comprising a single polypeptide chain. Very few monomeric enzymes are known and all of these catalyse a hydrolytic reaction. Monomeric enzymes typically comprise 100 to 300 amino acid residues and have molecular weights in the range 13 000 to 35 000.

Many proteolytic enzymes (proteases) are monomeric enzymes. These enzymes catalyse the hydrolysis of peptide bonds in other proteins and may be active against a variety of protein molecules. They are usually biosynthesised as larger inactive precursors known as **zymogens** (or **proenzymes**) to prevent generalised damage to cellular proteins. Zymogens are activated as required by the action of other proteolytic enzymes (also see Section 1.7).

1.4 Forces that stabilise protein molecules

The conformation of a protein is ultimately defined by its amino acid sequence, i.e. its primary structure. Upon biosynthesis of a polypeptide, folding is thought to occur simultaneously in many places, giving rise to secondary structure. The folded regions then interact to form structural motifs and domains and, finally, the domains associate to give rise to the tertiary structure of the protein, i.e. its native conformation. Whereas regions of secondary structure are stabilised by interactions between amino acids close together within the polypeptide chain, tertiary structure is stabilised by interactions between amino acids that are far apart on the chain but that are brought into close proximity by protein folding. The main forces that stabilise tertiary structure are hydrophobic interactions, electrostatic interactions and covalent linkages. Table 1.2 compares bond strengths of interactions involved in stabilising protein molecules.

Hydrophobic interactions arise through the tendency of non-polar hydrophobic amino acid residues, such as leucine and valine, to minimise contact with water molecules. The most energetically favourable way for a protein to fold is therefore to enclose the hydrophobic side chains of such amino acids in the interior of the protein, i.e. remove them from the

Table 1.2 Comparative bond strengths of interactions involved in stabilising protein molecules

Interaction type	Approximate bond energy ($kJ\,mol^{-1}$)
Disulfide bond	200
Ionic interactions	86
Hydrogen bond	10–20
Van der Waals forces	10
Hydrophobic interactions	0

surrounding aqueous (polar) environment. Hydrophobic interactions are the most important stabilising influence on protein tertiary structure.

Electrostatic interactions include ionic interactions, hydrogen bonds and Van der Waals forces. Ionic interactions (salt bridges) occur between oppositely charged amino acid side chains, e.g. between the amino group (—NH_3^-) of lysine and the carboxyl group (—COO^-) of glutamate. Ionic interactions, which are found mainly on the surface of folded polypeptides, contribute only modestly to tertiary structure. The strength of interaction is determined by the proximity of the groups and the medium that separates them, the force between the two charges being inversely proportional to the dielectric constant of the medium. Hydrogen bonds are stronger than hydrophobic interactions and occur extensively in proteins (Table 1.2). However, they do not contribute substantially to overall conformational stability of a polypeptide because in an unfolded state, hydrogen bonds would form with surrounding water molecules, i.e. there is little energy difference between folded and unfolded states. These bonds, however, do influence protein folding and, therefore, the native tertiary conformation, because they are strongest when the atoms involved are arranged linearly, e.g. in an anti-parallel β-sheet (Fig. 1.7c).

Van der Waals forces are relatively weak electrostatic interactions that occur between atoms and molecules when they are an optimum distance apart (Table 1.2). Electrons (negatively charged) of an atom become attracted to the nuclei (positively charged) of other atoms, giving rise to dipoles. Although each van der Waals interaction is weak, there are large numbers of interactions and overall they contribute significantly to tertiary protein structure.

Covalent bonds are far stronger than electrostatic interactions. Disulfide linkages are the major covalent bonds that help stabilise the tertiary structure of a protein and occur between cysteine residues. They are formed after the protein has folded into its native tertiary conformation and therefore stabilise the structure but do not direct folding. Disulfide bonds are rarely formed in intracellular proteins because of the reducing environment. They are formed in proteins exposed to more oxidising environments such as many extracellular proteins and outer membrane proteins. Disruption of disulfide bonds often has little effect on the tertiary structure, although in cysteine-rich polypeptides the protein tends to become less conformationally stable.

Polypeptide chains of oligomeric proteins are usually linked together by non-covalent interactions; peptide bonds are never involved.

1.5 Properties of proteins

The chemical properties of proteins are essentially those of the side chains of the amino acid residues. Several chemical reactions with proteins

Table 1.3 Some of the reactions that can be used to estimate the quantity of protein in solution

Reaction name	Reactants	Colour produced
Hopkins–Cole	Indole group of tryptophane / glyoxylic acid and sulfuric acid	Purple
Millon	Phenolic group of tyrosine / mercuric sulfate and sodium nitrate	Red
Folin–Ciocalteu	Phenolic group of tyrosine / tungstate and molybdate	Blue
Sakaguchi	Guanidine group of arginine / α-naphthol and oxidising agent	Red
Biuret	Peptide bonds / alkali cupric sulfate	Purple
Lowry	Combination of biuret and Folin–Ciocalteu reactions	Blue

produce coloured products, where the intensity of the colour produced is proportional to the number of reacting groups present (Table 1.3). Such reactions can be used for quantitative estimation of protein concentration in solutions. For many of these reactions it is necessary to assume that the protein being estimated has an average distribution of amino acid residues and that free amino acids (or other reacting groups) are not present. In the case of the Biuret reaction (and Lowry method), however, free amino acids do not react because it is based on reaction with the peptide bonds.

Although most proteins are not coloured, they all absorb light in the ultraviolet (UV) region of the spectrum. UV absorbance of proteins is particularly strong at 280 nm owing to the side chains of tryptophan and tyrosine. Measurement of protein absorbance at this wavelength can be used as a ready but rough estimation of protein in solution.

Functional groups in amino acid side chains have an important role in enzymatic catalysis. This is considered later in this chapter (Section 1.8) and in Chapter 3. Inactivation of enzymes can also occur by binding of chemical agents to these functional groups. For example, sulfydryl groups of cysteine bind heavy metal ions strongly, which can inactivate many enzymes:

$$\text{Enzyme—SH} + \text{Au}^+ \rightarrow \text{Enzyme—S—Au} + \text{H}^+$$

1.5.1 Acid–base properties

Proteins contain ionisable groups and therefore display acid–base properties. By definition, an acid is a proton donor and a base is a proton acceptor. Any acid–base reaction may be written as:

$$\text{HA} \Leftrightarrow \text{H}^+ + \text{A}^-$$

where HA is an acid and A^- is a base.

The strength of an acid depends on its dissociation constant, K_a. A strong acid dissociates readily, whereas a weak acid dissociates only partly (i.e. has a higher affinity for its proton). Generally, dissociation (K_a) of an acid may be described as:

$$K_a = \frac{[H^+][A^-]}{HA}$$

Rearranging:

$$[H^+] = \frac{K_a[HA]}{[A^-]}$$

Taking the negative logarithm of both sides:

$$-\log_{10}[H+] = -\log_{10} K_a + \log_{10}\frac{[A^-]}{[HA]}$$

Defining pH and pK_a:

$$pH = pK_a + \log_{10}\frac{[A^-]}{[HA]}$$

The final expression gives the relationship between pH and the degree of ionisation of the ionisable species [HA]. It is known as the Henderson–Hasselbalch equation.

All amino acids will be ionised in aqueous solution in a manner determined by pH. From Fig. 1.10 it can be seen that the pK_a values are the points of inflection on the curve. Two ionisation curves are shown, giving two pK_a values, because the α-carboxyl and α-amino groups of the free amino acid are ionised over different pH ranges. If several ionisable groups are present, as is the case for an enzyme, the situation is more complicated. In proteins, the ionisable groups contributing to the acid–base properties are the C-terminal α-carboxyl group, the N-terminal α-amino group and the side chains of certain other amino acids (arginine, aspartate, cysteine, glutamate, histidine, lysine and tyrosine). Other α-carboxyl and α-amino groups do not contribute to acid–base properties because they form the peptide bonds in the protein. Approximate pK_a values range from 3.0 for the α-carboxyl group, through 8.0 for the α-amino group to 12.5 for the guanidine group of arginine. The overall titration curve for a protein will comprise the effects of all ionisable groups present in the molecule.

The titration curve for an amino acid shown in Fig. 1.10 shows that the buffering capacity of an ionisable group is greatest where the pH is near its pK_a value. Globular (soluble) proteins act as buffers that have the greatest buffering capacity near the pK_a value of the most frequently occurring ionisable group in the molecule.

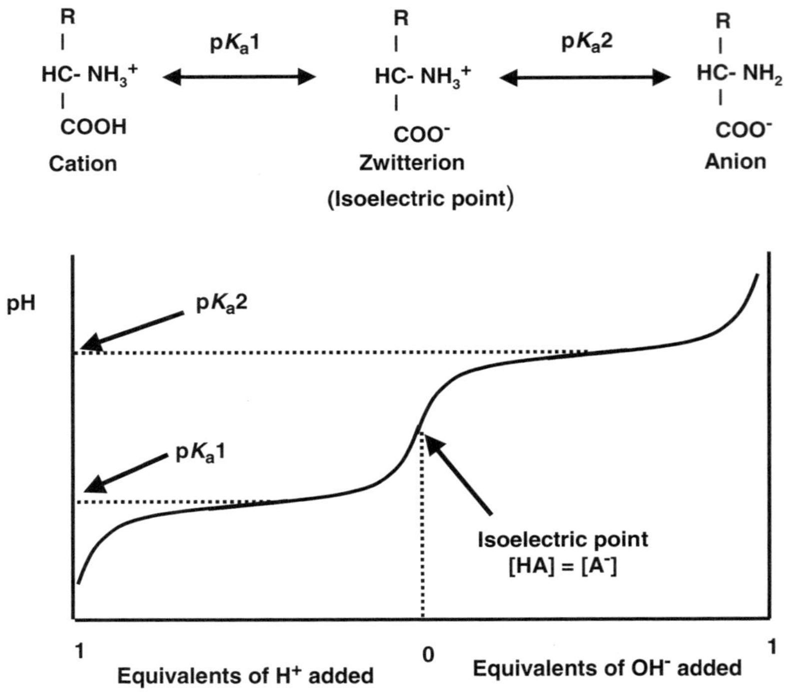

1.10 Titration curve for an amino acid.

Just as the charge on a free amino acid is determined by pH, as shown in Fig. 1.10, so is that on a protein. At low pH most proteins carry a net positive charge, whereas at high pH most proteins carry a net negative charge. The pH at which there is an equal balance between negative and positive charge (i.e. no net charge on the protein) is known as the isoelectric point.

Enzyme functioning is pH dependent and each enzyme has a characteristic pH at which catalytic activity is highest (optimum pH), as well as minimum and maximum values beyond which the enzyme does not function. Most enzymes are active within the pH range 5 to 9 and many display a bell-shaped curve of activity against pH, as illustrated in Fig. 1.11a. The influence of pH on enzyme activity is a result of a combination of (1) the ionisation state of the amino acid side chains involved in the catalytic activity of the enzymes and (2) the ionisation of the substrate. Binding of substrate to the enzyme lysozyme, for example, is related to the ionisable state of two key amino acids: asparagine at position 52 and glutamate at position 35. The enzyme catalyses hydrolysis of the major cell wall polyscaccharides of bacterial cells. At the pH optimum for lysozyme (pH 5), asp-52 is ionised ($—COO^-$), while glu-35 is not ($—COOH$). The ionisable state of these two

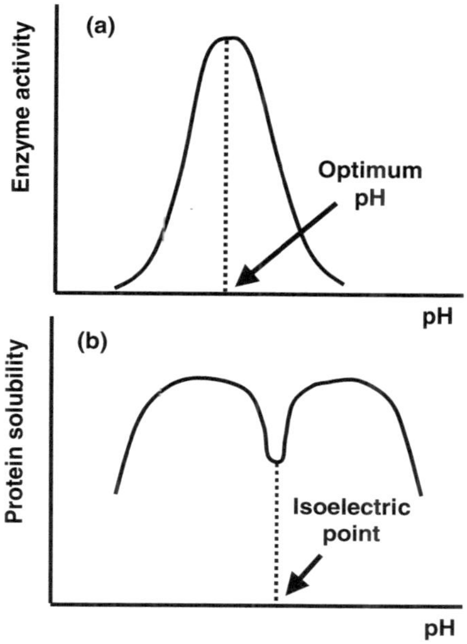

1.11 Typical profiles for influence of pH on (a) enzyme activity and (b) protein solubility.

amino acids has a strong influence on substrate binding and thus enzyme activity. Activity decreases rapidly below the pH optimum because asp-52 becomes protonated (—COOH). At pH values above the optimum, activity decreases because glu-35 becomes ionised ($—COO^-$). At extremes of pH, the charge on the ionisable side chains causes the compact tertiary structure (see Section 1.3.5) to disrupt and the protein denatures. For soluble (globular) proteins, denaturation leads to a marked reduction in solubility since hydrophobic groups normally on the inside of the molecule become exposed to the aqueous solvent. The solubility of an enzyme also decreases around its isoelectric point (Fig. 1.11b). The mechanism for this is formation of insoluble aggregates of the neutral molecules at the isoelectric point which, at other pH values, would have been repelled by the net charge they carry.

1.5.2 Other factors influencing protein solubility

Temperature and **salt concentration** are other important factors influencing the solubility of enzymes in aqueous solution. Enzyme activity increases with temperature up to an optimum temperature beyond which activity decreases sharply as the tertiary structure is disrupted through thermal

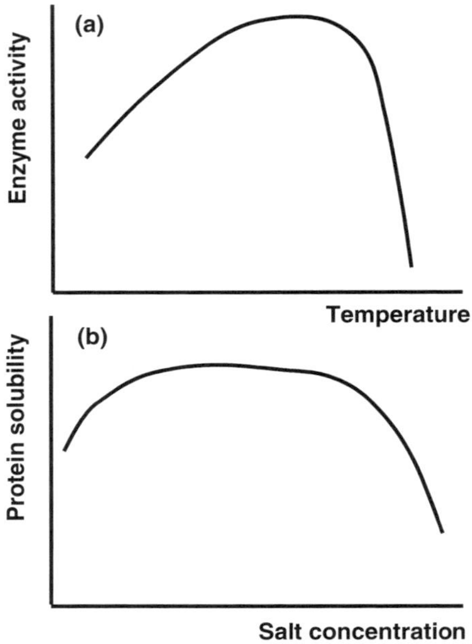

1.12 Typical profile for influence of (a) temperature on enzyme activity and (b) salt concentration on protein solubility.

agitation leading to denaturation (Fig. 1.12a). Change in solubility of enzymes with temperature largely mirrors that for change in activity.

At low salt concentration, the solubility of an enzyme increases owing to changes in ionisation of amino acid side chains. At very high salt concentrations, however, the high level of interaction between the added ions and aqueous solvent reduces the interaction between the protein and water, leading to precipitation of the enzyme from solution (Fig. 1.12b).

1.6 Biosynthesis of proteins

All cells have deoxyribose nucleic acid (DNA) as their genetic material, which ultimately codes for all the components of the cell including its enzymes. There are two essential processes that convert the information in DNA into protein. Transcription produces a ribose nucleic acid (RNA) copy, or messenger molecule, which is then translated into protein. These processes form the so-called central dogma of modern biology (proposed by Crick in 1956), which proposes a one-way flow of genetic information from DNA, through RNA to protein:

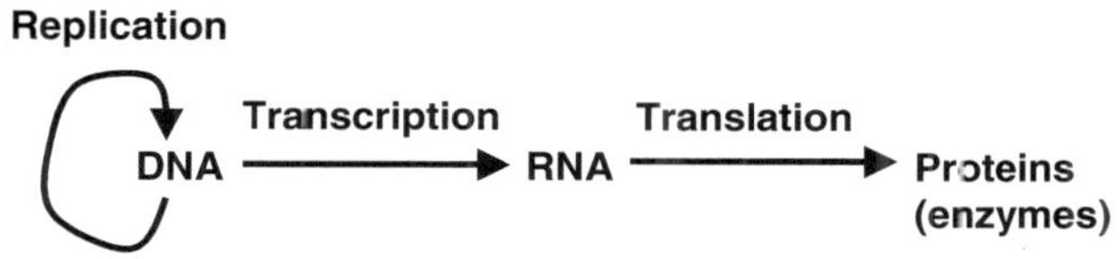

i.e. DNA makes RNA makes protein

The central dogma remains essentially unchanged today, although some rare exceptions to the rule of one-way flow of information from DNA to RNA are known. A detailed consideration of the central dogma is beyond the scope of this chapter. This section will explain, in broad terms, the process by which the sequence of nucleotides in RNA is converted into a sequence of amino acids in proteins – the process termed **translation**. A basic knowledge of the structure of DNA including its base pairing (i.e. A-T and C-G) is assumed. As the central dogma indicates, the base sequence of DNA indirectly specifies the amino acid sequence in proteins, i.e. an intermediate RNA molecule is involved that directly specifies the primary polypeptide structure. Other genes in DNA code for RNA molecules with a different function (see below). The enzymes that transcribe DNA to RNA are called RNA polymerases. Unlike DNA, RNA is usually a single strand of polynucleotide because it is formed by copying the base sequence of only one of the strands of DNA. In RNA, uracil replaces the thymine base in DNA. The RNA code therefore comprises adenine (A), uracil (U), guanine (G) and cytosine (C). Three major types of RNA are involved in protein synthesis: messenger RNA (mRNA), transfer RNA (tRNA) and ribosomal RNA (rRNA).

For protein synthesis to occur, the base sequence of DNA must be transcribed by RNA polymerase II to produce a mRNA molecule (the process of translation). It is the base sequence of the mRNA that codes directly for the amino acid sequence of the protein. Each amino acid is coded for by a triplet of bases (known as a codon). Since there are four different bases in RNA, there are 4^3 (=64) different triplet codons. It follows that since there are only 20 amino acids found in proteins, many must be coded for by more than one codon. The sequence of nucleotides in a codon that specifies a given amino acid is known as the genetic code and is shown in Fig. 1.13. Because more than one codon can determine the same amino acid the code is said to be 'degenerative'. Where degeneracy occurs it tends to occur in the third nucleotide of the codon with U and C and also A and G being equivalent to each other. Serine, for example, can be coded for by UCU, UCC, UCA or UCG. Degeneracy has the advantage that a mutation held in the third nucleotide has a lower chance of modifying the code than a

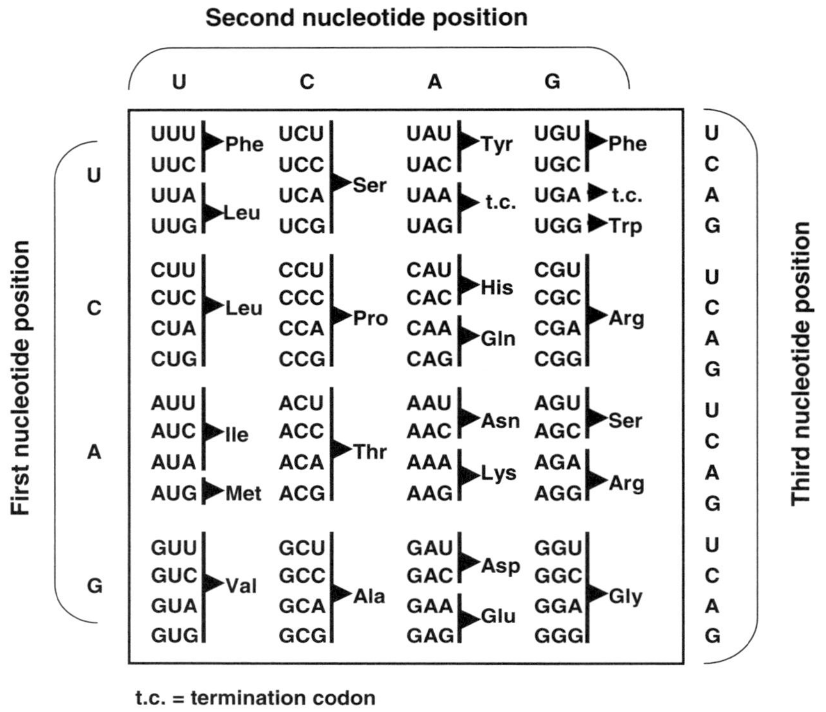

1.13 Genetic code: the sequence of nucleotides in a codon that specifies a given amino acid.

change in the other two nucleotides of the codon. As shown in Fig. 1.13, the code is not ambiguous as no codon codes for more than one amino acid.

Codon triplets of nucleotide bases cannot themselves recognise amino acid molecules. For protein synthesis, adaptor molecules are therefore required to align up amino acids on adjacent mRNA codons. The adaptors are tRNA molecules that are linked to their particular amino acid by highly specific enzymes (amino-acyl tRNA synthetases). There are at least 20 different types of these enzymes per cell: one for each amino acid. tRNA molecules are transcribed from DNA by RNA polymerase III enzymes. Each tRNA has a complementary sequence of bases (known as an anticodon) for the codon for its amino acid, which recognises the correct codon on the mRNA. The amino acids of the poypeptide are therefore arranged in the sequence defined by the order of codons in the mRNA. Linking an amino acid to its tRNA has the effect of activating it for participation in protein synthesis. The high specificity of the enzyme for the amino acid and the tRNA to which it is joined is essential for reliable protein synthesis. Subsequent translation of the mRNA code to give the polypeptide occurs on particles in the cytoplasm called ribosomes. These particles are composed of proteins and rRNA. Ribosomes from both prokaryotic and eukaryotic cells are made up of large and small subunits. They have two grooves: one accommodates around 35 nucleotides of the mRNA molecule, the other accommodates the growing polypeptide chain. Ribosomes also have two binding sites for tRNA molecules, one for peptide (termed P) and one for amino acid (termed A). The orientation of the P and A sites are such that for a tRNA molecule to bind, its anticodon must be complementary to the codon on the mRNA.

Figure 1.14 illustrates the process of translation of the message of mRNA into a polypeptide chain or protein. Once the P-site is occupied, an aminoacyl tRNA binds to the vacant A-site. The peptide chain grows by the carboxyl end becoming detached from the P-site tRNA and forming a peptide bond with the amino end of the amino acid at the A-site tRNA. The tRNA then leaves the P-site and the peptide-tRNA on the A-site moves to occupy the vacant site. It is the ribosome that moves in this process, not the tRNA. The movement involves a conformational change in one of the ribosomal proteins and is an energy-requiring step. Once the A-site is vacant the process starts again, involving binding of another aminoacyl tRNA. Each cycle takes only a twentieth of a second and bacteria can synthesise protein containing 300 amino acid residues in about 20 seconds.

Protein synthesis (translation) has three main stages: initiation, elongation and termination. Translation in eukaryotic cells is initiated by a special tRNA (initiator tRNA) recognising the start code of protein synthesis; the codon AUG. The initiator tRNA carries a methionine residue (tRNA-met) and is specific for the start codon. The initiator binds with the small ribosome subunit, in a process also involving initiation factors, to form a complex (Fig. 1.14a). The tRNA-met-small subunit-initiation factor complex associates with a mRNA molecule. Binding of tRNA-met with the start codon AUG then occurs and the initiation factors are released. Only then does the large ribosomal subunit bind to form the complete ribosome, with tRNA-met on the P-site. Aminoacyl tRNA can then bind to the A-site and protein synthesis proceeds (Fig. 1.14a). The initiation process so described is similar in prokaryotic cells, except that the initiator tRNA is linked to *N*-formyl methionine and not methionine.

In the elongation stage of protein synthesis, the formation of the peptide bond is catalysed by the enzyme peptidyl transferase (Fig. 1.14b). Addition of amino acids to the growing peptide chain continues until a stop codon in the mRNA is in the A-site of the ribosome (Fig. 1.14c,d). This leads to termination, in which a protein release factor binds to the stop codon and causes peptidyl transferase to hydrolyse the peptide-tRNA link (i.e. water molecule added instead of an amino acid). The polypeptide moves away from the ribosome and spontaneously coils into its secondary and tertiary

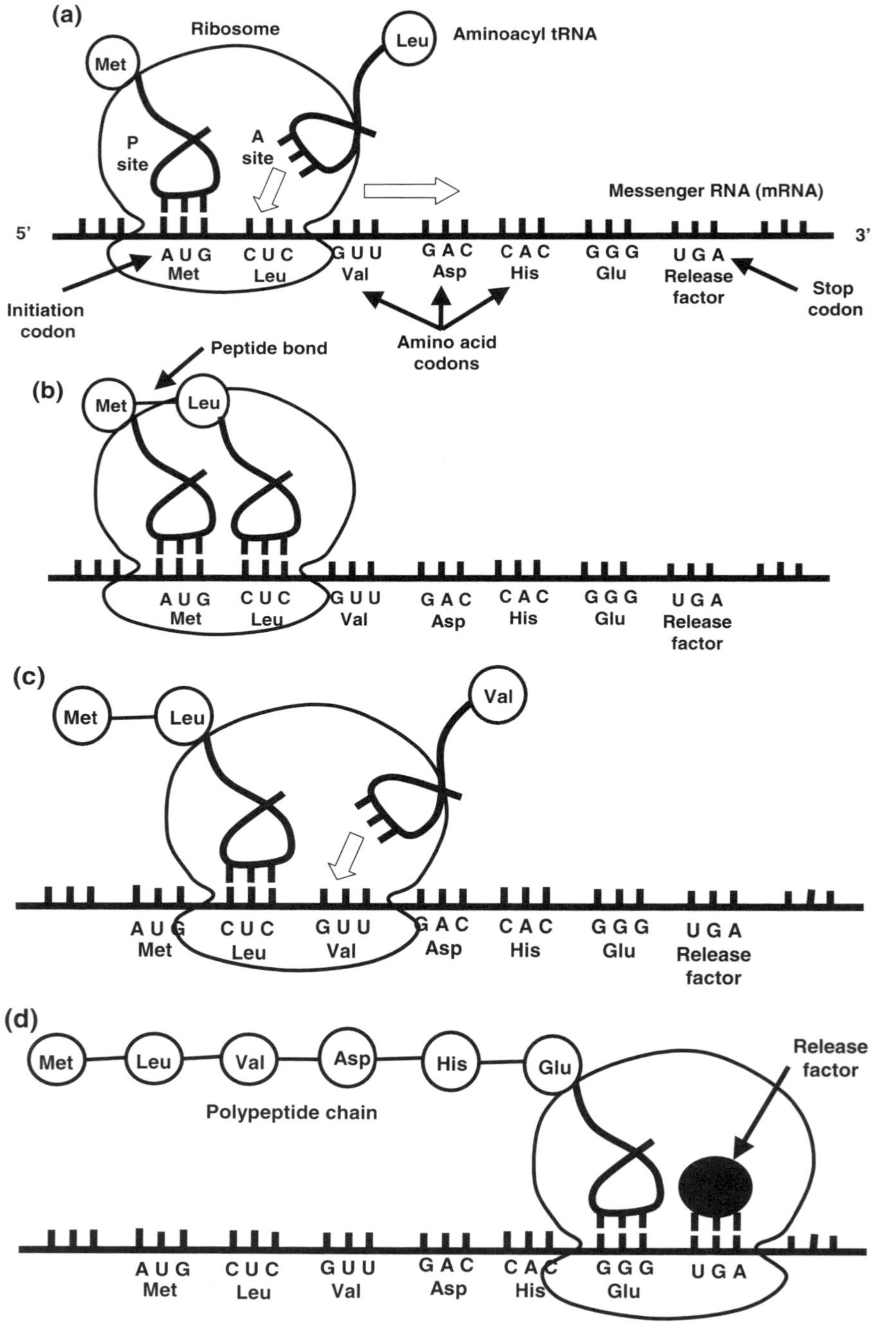

1.14 Protein biosynthesis in eukaryotes. (a) Initiation of protein biosynthesis involving binding of tRNA-met to the P-site and of an aminoacyl tRNA to the A-site. (b) Peptide bond formation catalysed by peptidyl transferase. (c) Elongation of the polypeptide chain. (d) Termination of protein synthesis involving binding of a protein release factor.

structure. The functional ribosome then disassembles, leading to separation of small and large subunits, tRNA and release factor.

The addition of an amino acid to a growing polypeptide chain is energy demanding, requiring four high-energy bonds to make a new peptide bond; protein synthesis is the most energy demanding of all cellular biosynthetic processes.

1.7 Post-translational modification of proteins

Post-translational modification refers to the covalent modification of many proteins that occurs during or after their formation on ribosomes. Phosphorylation, glycosylation and proteolytic processing are the most common forms of post-translational modification, although several others are known to occur. These modifications tend to affect the structural stability or biological activity of the protein.

Reversible **phosphorylation** of proteins occurs through the action of specific protein kinase and protein phosphatase enzymes, giving rise to phosphorylated and dephosphorylated forms of the protein, respectively. Adenosine triphosphate (ATP) is the usual phosphate group donor. The phosphate group is transferred to the hydroxyl group of serine, threonine or tyrosine residues, although the side chain of certain other amino acids can also be phosphorylated. In most cases, phosphorylation–dephosphorylation functions as a reversible on-off switch with respect to biological activity. Some proteins are biologically activated by phosphorylation, whereas others are inactivated. Phosphorylation also alters the physicochemical properties of a protein through increase in the number of negative charges carried.

Glycosylation refers to the attachment of carbohydrates. It is a common post-translational modification of extracellular and cell surface proteins of eukaryotic cell origin. There are two types of glycosylation: N-linked and O-linked. In N-linked glycosylation the oligosaccharide (carbohydrate chain) is linked to the nitrogen atom of an asparagine residue, whereas O-linked glycosylation involves oligosaccharide linkage to the oxygen atom of hydroxyl groups (usually of serine or threonine residues). The oligosaccharide component of a glycoprotein has various potential functions: direct role in the biological activity of the protein; targeting of proteins to specific cellular locations; helping stabilise the protein; protecting from proteolytic attack; enhancing solubilisation of protein; increasing biological half-life of protein; enhancing recognition and cell–cell adhesion.

Proteolytic post-translational modification occurs for some proteins. Generally, these proteins are inactive when synthesised (the 'pro' form or precleaved state) with functional activation occurring upon proteolysis; also see Section 1.3.7. Such proteolytic cleavage is very specific and generally irreversible. Most of the proteins that are processed in this way are ultimately exported to cellular organelles or secreted from the cell.

Acetylation (CH_3CO—group transfer) of proteins can occur during synthesis or post-translationally. The enzyme involved is *N*-acetyl transferase, which is loosely associated with ribosomes and usually uses acetyl-CoA as acetyl donor. More than half of all polypeptides synthesised in the cytoplasm have an N-terminal acetyl group. The function of N-terminal acetylation and the criteria governing selection of polypeptides for this modification are not fully understood.

Protein **acylation** is thought to occur in all eukaryotic cells, and a wide range of cytoplasmic and membrane proteins are modified through acylation. In this process, polypeptides are modified by direct covalent attachment of fatty acids. The saturated fatty acids, palmitic acid (C_{16}) and myristic acid (C_{14}) are most commonly involved. Palmitic acid is attached through ester or thioester bond formation to either a cysteine, serine or threonine residue. Several enzymes are thought to be involved in this type of post-translational protein modification. Myristic acid is always covalently attached to proteins via an amide bond to an N-terminal glycine residue. This modification occurs before polypeptide synthesis is complete and is catalysed by the enzyme myristoyl CoA:protein *N*-myristoyl transferase. Acylation is thought to play a role in interaction of proteins with biological membranes or with other proteins. Many acylated polypeptides are found in multi-subunit protein complexes and removal of the fatty acid component has a negative affect on subunit interaction.

1.8 Enzymatic catalysis

Enzymes are catalysts – they speed up the rate of chemical reactions but remain unaltered themselves.

Chemical reactions proceed when the free energy of the products is less than that of the reactants. In spontaneous reactions therefore, the change in free energy (ΔG) is positive, i.e. energy is released. Although biochemical reactions of cellular metabolism are spontaneous in the sense that they release energy, they proceed exceedingly slowly in the absence of enzymes because of the energy 'barrier' between substrate(s) and product(s). This barrier is known as the 'activation energy' and for molecules to react, they must possess energy corresponding to the top of this barrier. Heating a solution containing substrate molecules is one way of increasing the energy levels. Enzymes, however, do not increase energy levels of substrate molecules; rather they provide an alternative low-energy route for the reaction to proceed. Enzymes therefore function as catalysts by lowering the activation energy of a reaction. The effect is to accelerate spontaneous (energy releasing) reactions. Essentially, more molecules of substrate have sufficient energy to react in an enzyme-catalysed reaction. Some enzymes have enormous catalytic power. For example, catalase which catalyses the breakdown of hydrogen peroxide ($2H_2O_2 \rightarrow 2H_2O + O_2$) accelerates the reaction around 10^{14} times (a million billion) compared to the uncatalysed reaction, each molecule of catalase being capable of processing around five million molecules of hydrogen peroxide per second!

Enzyme catalysis involves formation of a transition state (ES) in which the substrate is bound to the enzyme, i.e. $E + S \leftrightarrow ES \rightarrow E + P$. Generally, anything that promotes the formation of the transition state will enhance the rate of reaction.

Mechanisms of enzyme catalysis are considered in Chapter 3.

1.8.1 Specificity

Enzyme specificity depends on two important aspects:

- What molecule(s) any particular enzyme acts on;
- What reaction is performed.

All enzymes can discriminate between different molecules but the extent of discrimination varies between different enzymes. The enzyme urease, for example, is highly specific, being capable of acting only on urea (H_2NCONH_2) and not on closely related molecules (e.g. $H_2NCONHCH_3$). Other enzymes can act on closely related molecules. The proteolytic enzyme trypsin, for example, hydrolyses the peptide bond to the C-terminal side of either arginine or lysine. Trypsin is most active at pH 8.0, at which arginine or lysine are the only amino acids with side chains that are positively charged. The enzyme therefore tolerates minor differences in the shape of its substrate, provided that the amino acid side chain is positively charged.

Most enzymes are highly specific with regard to which groups and bonds are acted upon, and also therefore what products are formed. This positional or regional specificity (regiospecificity) is vital to cellular metabolism. The enzyme hexokinase that catalyses the initial phosphorylation of glucose in metabolism displays absolute positional specificity; although glucose has five hydroxyl groups available for phosphorylation, the enzyme always adds the phosphate group at the same C6 position.

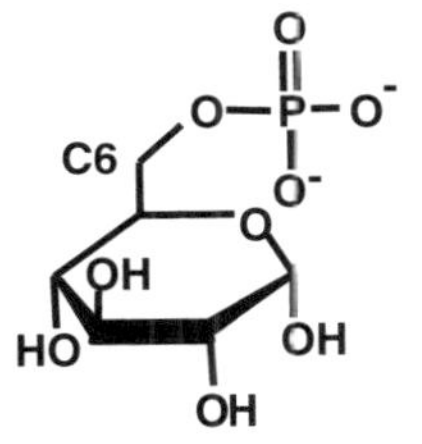

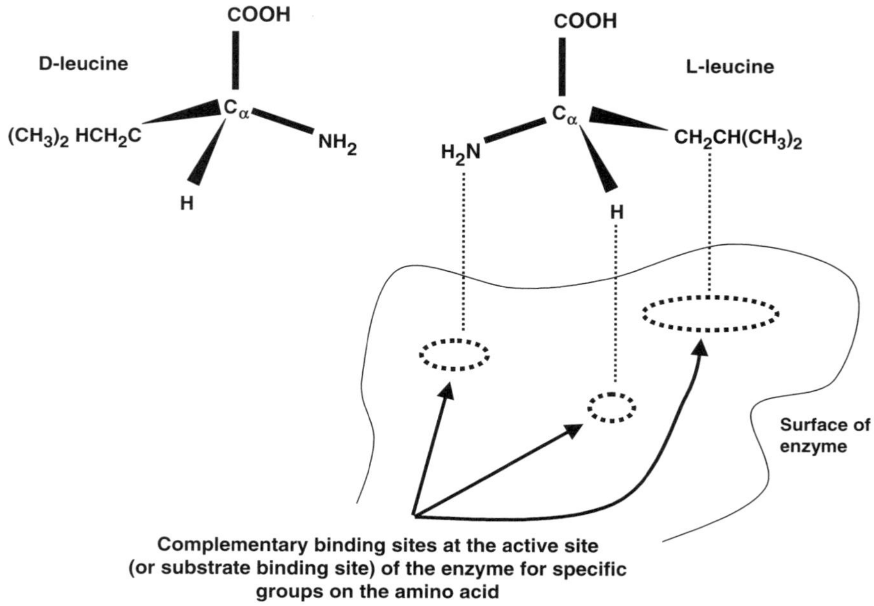

1.15 Enzyme discrimination between stereoisomers.

Enzymes also display high specificity with regards to stereoisomers. Enzymes that act on optically active compounds normally show absolute specificity for either the D- or the L-isomer. For example, hexokinase is active on D-glucose and not on L-glucose, whereas propteolytic enzymes are active towards polypeptides made up exclusively of L-amino acids. As indicated earlier in this chapter (Section 1.3.2), sterioisomers arise in organic molecules when an asymmetric centre occurs, i.e. there is a carbon bonded to four different atoms or groups. The only difference between the pairs of isomers is the distribution of the substituent atoms or groups around the centre carbon atom. In free solution these isomers will react in the same manner. However, when bounded to an enzyme with complementary sites for three of these groups, the position of the binding sites dictates which of the two isomers binds and then reacts. This is illustrated in Fig. 1.15 for an amino acid leucine, where only L-leucine binds to the enzyme. Although binding of substrate to the enzyme might confirm absolute stereospecificity, it might not necessarily be the case. Isomers would not be distinguished by an enzyme that required only two binding sites to be complementary, rather than three.

Some enzymes are far less specific and there are circumstances in which absolute specificity is not desirable to organisms. Hydrolytic enzymes, including many proteases, tend to display a relative lack of specificity. These enzymes are often involved in breakdown of large molecules, releasing component parts that may be utilised as a growth nutrient. Chymotrypsin, for example, cleaves polypeptides that have an amino acid with an aromatic or other hydrophobic side chain on the carbonyl side of the peptide bond. This enzyme can also act on small ester substrates, such as *p*-nitrophenyl acetate:

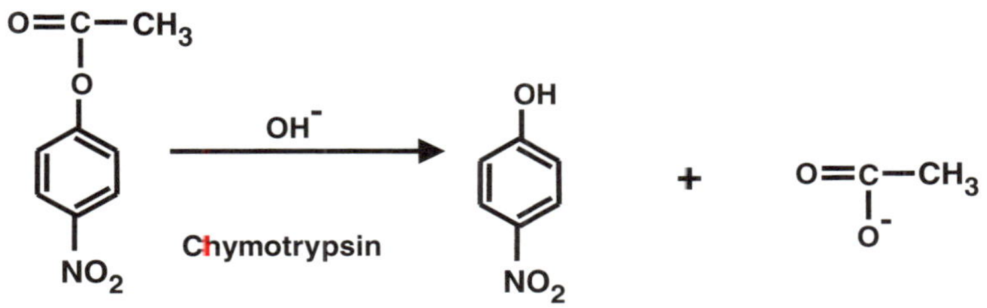

The enzyme, whether acting on *p*-nitrophenyl acetate or polypeptides, acts as an esterase.

Enzymes distinguish between different molecules by the particular shape and charge distribution of the active site of the enzyme. The active site is the region that binds the substrate and is only a small part of an enzyme. For some enzymes, as few as 5–6 amino acid residues within the protein form the active site. The active site must ensure not only that the substrate binds, but also that other molecules are excluded and that the substrate is in the correct orientation for catalysis to take place.

Fischer, as far back as 1890, proposed that the shape of the active site of an enzyme was complementary to that of its substrate and that this explained the specificity of enzyme action. This became known as the 'lock and key' model of enzyme–substrate interaction. The model provides a convenient explanation not only of the absolute specificity of some enzymes, but also of the more relaxed specificity of others (Fig. 1.16a). Enzymes, however, are now known to be flexible molecules and the lock and key model implies a rigid structure. Koshland, in 1958, proposed an alternative model for enzyme action in which the enzyme itself undergoes conformational change as a result of substrate binding (Fig. 1.16b). In this 'induced fit' model, the substrate induces a change in the shape of the active site to the correct fit. This occurs only when the substrate binds, which implies that the shape of the active site and its substrate are complementary only when in a transitional state of an enzyme–substrate complex. An enzyme may also distort (or strain) the substrate (Fig. 1.16c).

Enzymes bind to substrates by a combination of hydrogen and ionic bonds, as well as by hydrophobic and van der Waals interactions. Covalent

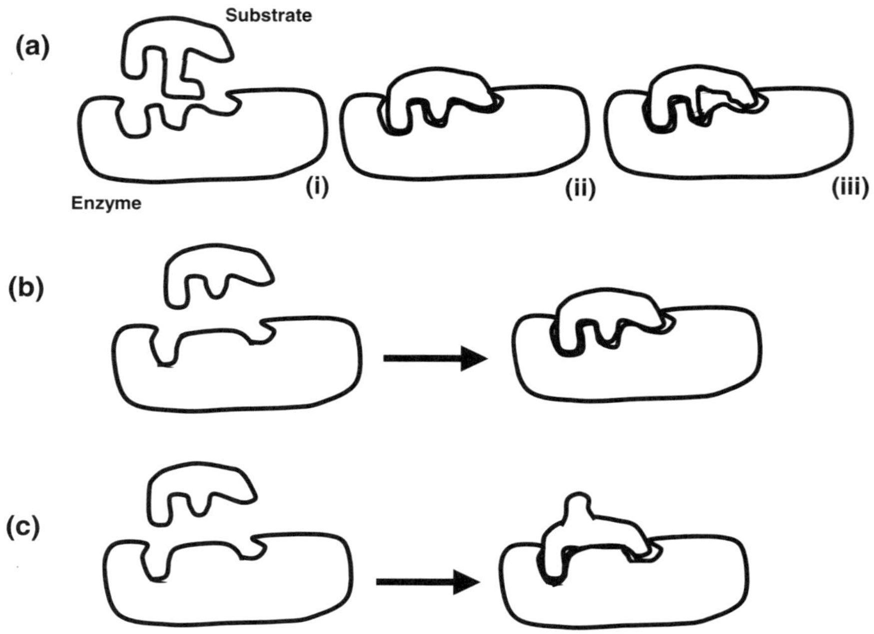

1.16 Models of enzyme active site interaction with substrate. (a) The 'lock and key' model for an enzyme active site with relaxed specificity: (i) no binding of incompatible substrate; (ii) binding of substrate having high complementarity with active site; (iii) binding of substrate having low complementarity with active site. (b) The 'induced fit' model in which the substrate induces a change in the shape of the active site to the correct fit. (c) Deformation (or straining) of substrate leading to improved fit with active site.

bonds are only occasionally formed and then usually as part of the catalyst. The strength of binding is a function of the distance apart of atoms; repulsion occurs if too close and increasing distance results in a progressively weaker interaction. Generally, strong binding results from complementarity of both shape and charge for the active site and its substrate. This is illustrated in Fig. 1.17 for a proteolytic enzyme capable of binding both alanine and lysine. The active site makes very limited contact with the side chain of alanine, resulting in low strength binding and a low tendency to bind. For lysine, however, the side chain makes numerous close contacts (van der Waals interactions) with the active site and an ionic bond is formed at the bottom of the active site; both the strength and tendency of lysine binding are relatively high.

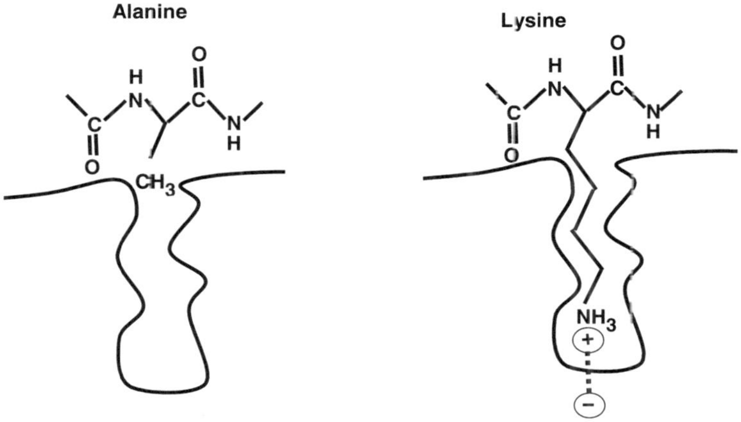

1.17 Complementarity of substrate and active site of an enzyme governs strength of binding.

1.8.2 Enzyme cofactors

The catalytic activity of some enzymes requires the presence of non-proteinaceous compounds. Such compounds are collectively known as cofactors. Enzymes that require cofactors are often commonly referred to **apoenzymes** when not bound to the cofactor and are catalytically inactive, and as **holoenzymes** when bound to the cofactor and catalytically active (Fig. 1.18).

Cofactors are broadly categorised into three types:

1. Prosthetic groups are organic cofactors that are tightly bound to the enzyme, sometimes covalently. Flavin adenine dinucleotide (FAD) of some dehydrogenase enzymes and biotin of some carboxylase enzymes are examples of prosthetic groups (Fig. 1.19).
2. Coenzymes are organic cofactors that are more easily removed from the enzyme than prosthetic groups. Examples are nicotinamine adenine dinucleotide (NAD^+) or nicotinamine adenine dinucleotide phosphate ($NADP^+$), which are required by many dehydrogenase enzymes for catalytic activity (Fig. 1.19). NAD^+ and $NADP^+$ accept reducing equivalents from a variety of substrates depending on the particular enzyme. For example:

$$CH_3CH_2OH + NAD^+ \xrightarrow[\text{dehydrogenase}]{\text{Alcohol}} CH_3CHO + NADH + H^+$$

Coenzymes can often be regarded as a second substrate for the enzyme. The dehydrogenase enzymes, for example, have a strong binding site for the oxidised form of their coenzyme (NAD^+). Once the substrate is

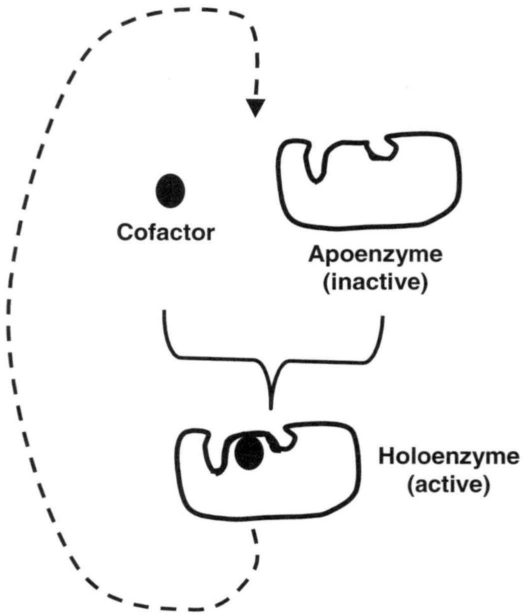

1.18 Cofactor interaction with an enzyme. Some cofactors (mostly coenzymes) are not bound tightly to enzymes and are readily lost when enzymes are extracted from cells, leading to inactive forms of the enzymes. Enzyme activity is normally restored by addition of the essential cofactor.

oxidised, the reduced form (NADH) leaves the enzyme and is reoxidised in cellular metabolism. The NAD^+ so formed is available for binding to the dehydrogenase enzyme and the cycle is repeated. In this way, NAD^+ acts as a second substrate for the enzyme, but unlike most substrates it must be continually recycled within the cell.

3. Metal ions are cofactors for some enzymes. They may be loosely or tightly bound and, in some cases, are associated with the prosthetic group (e.g. Fe^{2+} or Fe^{3+} in the heme group). These ions are similar to coenzymes in the sense that they confer on the enzyme a property it would not possess in their absence. In other cases, free metal cations such as Fe^{3+}, Zn^{2+}, Cu^{2+} and Mg^{2+} function as cofactors.

Table 1.4 gives examples of cofactors and their catalytic roles.

1.9 Future trends

Technical advances in NMR and computational analysis of spectra are likely to enable structural determination of relatively high molecular weight

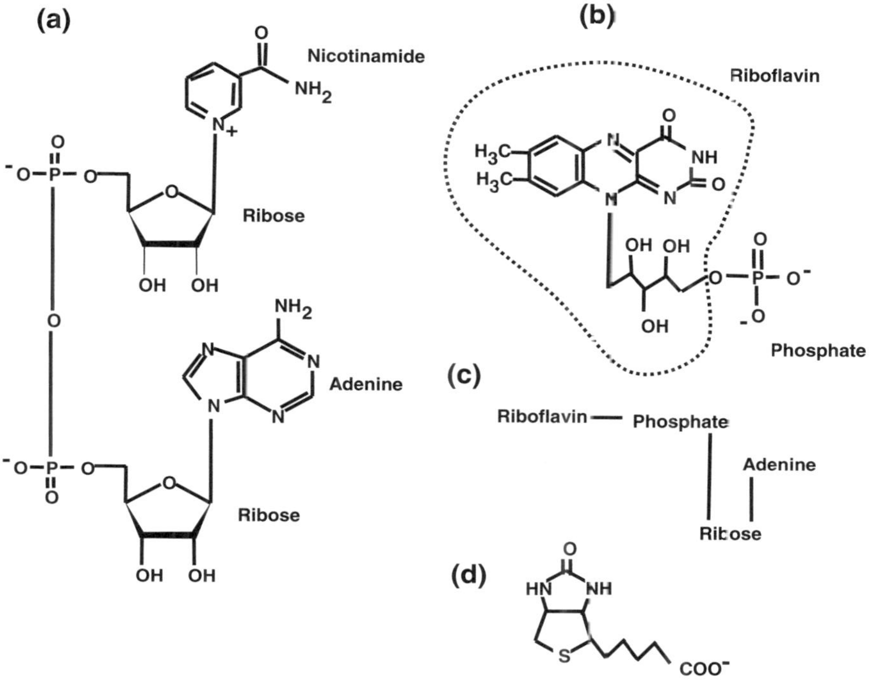

1.19 Structures of some important cofactors. (a) Nicotinamide adenine dinucleotide (NAD^+). $NADP^+$ differs from NAD^+ only by the presence of a phosphate group on the adenine. (b) Flavin mononucleotide (FMN). (c) Flavin adenine dinucleotide (FAD). (d) Biotin.

proteins in solution. Fourier Transform Infrared Spectrometry (FTIR) is also likely to contribute strongly to analysis of protein secondary structure in the future. The advantage of FTIR over other techniques is that spectra can be obtained for proteins in a wide range of environments, for example in solution, on various surfaces. Improved characterisation of infrared spectra for proteins with defined secondary structure content (using synthetic polypeptides), together with advances in computational analysis, is providing insights into the frequency and variability of particular secondary structural conformations. FTIR continues to offers great potential for investigating the influence of physical and chemical factors, such as temperature and pH, on the secondary structure of proteins in solution.

Through site-directed mutagenesis (or protein engineering) it is now possible to manipulate the amino acid sequence of a protein. The technique, which involves use of synthetic DNA fragments to change the genetic information (DNA) coding for a protein, an improved understanding of the

Table 1.4 Examples of enzyme cofactors and their roles

Cofactor	Role
Nicotinamide adenine dinucleotide (NAD^+, $NADP^+$)	Oxidation/reduction reactions [those involving transfer of H^- (hydride) ion]
Flavin adenine dinucleotide (FAD)	Oxidation/reduction reactions
Flavin mononucleotides (FMN)	Oxidation/reduction reactions
Biotin	Carboxyl group transfer
Cobalamin	Methyl group transfer
Coenzyme A (CoA)	Transfer of groups, e.g. acetyl
Thiamine pyrophosphate (TPP)	Acetaldehyde transfer
Tetrahydrofolate (THF)	One-carbon transfer reactions
Pyridoxal phosphate	Transamination and decarboxylation reactions
Metal ions:	
Fe, Cu, Mo	Oxidation/reduction reactions
Zn	Helps bind NAD^+
Co	Part of cobalamin coenzyme
Mn	Aids in catalysis by electron withdrawal

relationship between the primary amino acid sequence and its higher structural conformation can be obtained. Site-directed mutagenesis is also a valuable tool for studying structural and functional relationships of enzymes and for enhancing the attributes for commercially important enzymes. The catalytic capabilities of several enzymes have been improved by changes to the amino acids forming the active sites. Other studies have shown that it is possible to improve temperature and pH stability of enzymes. For example, introduction of cysteine residues into the polypeptide chain tends to enhance protein stability through disulfide bond formation. Replacement of lysine with arginine in a polypeptide chain also tends to enhance enzyme stability through increase in the extent of overall hydrogen bonding. Ultimately, continuing advances in this field could facilitate routine *de-novo* design of enzymes to suite particular applications; the 'Holy Grail' of applied enzymology!

The advances made in molecular biology and the associated research equipment have accelerated DNA sequence determination and large portions of the genome of several species have been sequenced. This genomic information will increasingly drive future trends in enzymology – within the emerging fields of **proteomics** and **bioinformatics** – leading to greatly improved understanding of structure, function and expression of newly discovered proteins.

1.10 Further reading

Modern textbooks of biochemistry[1–3] provide a wealth of fundamental information on enzymes and proteins in a well-illustrated and readily digestible form. Typically such texts have sections dealing with protein structure, enzymes as biological catalysts and protein expression in biological systems. They are an excellent source of background reading on these topics, and are recommended for improving understanding of fundamental biological aspects. These texts are also recommended for broader reading into modern molecular methods of DNA technology.

Two books[4,5] devoted solely to 'enzymes' are recommended for more in-depth studies. A notable strength of both texts is the simplified and effective approach to the complex mathematical treatment of enzyme kinetic data. There is also a useful survey[4] of relevant Internet sites and computer software for enzymatic data analysis. Palmer's text[5] aims to provide a reasonably detailed account of all the various theoretical and applied aspects of enzymology likely to be included in a course on this subject – as a master educator, he delivers the material in a readily understandable form.

A book by Walsh[6] is also recommended as a source of information on industrial uses of proteins. It provides a broad overview of the various industrial uses of proteins, as well as an understanding of the fundamental biochemistry and methodological approaches that underpin such applications. The most notable feature that distinguishes it from standard molecular biology, biochemistry or biotechnology texts is the wide range of proteins considered and the depth of information provided. Sources of proteins are considered in some detail, including heterologous proteins from various microorganisms, plant and animal tissues and insect cells. In dealing with protein purification and characterisation, the theory is integrated well with the practical aspects and there is useful advice on scale-up of procedures. There is a chapter devoted to the various general issues relating to large-scale protein purification, with particular emphasis on the problems of contamination. Specific examples of how various proteins are produced commercially are also included in many of the subsequent chapters of the book. Chapters on industrial enzymes include – amongst others – proteases, carbohydrases, lignocellulases, pectinases, lipases and oxidoreductases. There is also an impressive collection of recent books and articles, with a good balance of general and specific sources of information.

A selection of review articles is provided[7–18] for those readers wishing to delve into a detailed consideration of particular aspects of protein and enzyme technology. These reviews are effective interfaces with the primary literature, which lead to key papers on the individual topics.

There are, of course, a host of Internet sites that provide information on proteins and enzymes, most of which provide useful links to other related sites. Structural computational analysis of proteins and detailed pictorial representations of proteins are notable strengths of the Internet in this context. A good starting point is The ExPASy (**Ex**pert **P**rotein **A**nalysis **Sy**stem) proteomics server of the Swiss Institute of Bioinformatics,[19] which is dedicated to the analysis of protein sequences and structures (see Reference section). A short list of other sites recommended of initial review is provided in the bibliography of this chapter.

1.11 Bibliography

1.11.1 Books

1. Garratt R H and Grishham C M (1999) *Biochemistry*, 2nd edition, New York, Saunders College Publishing.
2. Voet D and Voet J G (1999) *Fundamentals of Biochemistry*, Chichester, UK, John Wiley & Sons.
3. Stryer L (2002) *Biochemistry*, 5th edition, New York, W H Freeman.
4. Copeland R A (2000) *Enzymes: a practical introduction to structure, mechanism, and data analysis*, 2nd Edition, Chichester, UK, John Wiley & Sons.
5. Palmer T (2001) *Enzymes: biochemistry, biotechnology, clinical chemistry*, Chichester, UK, Horwood Publishing.
6. Walsh G (2002) *Proteins: biochemistry and biotechnology*, Chichester, UK, John Wiley & Sons.

1.11.2 Articles

7. Al-Lazikani B, Jung J, Xiang Z X and Honig B (2001) 'Protein structure prediction', *Curr. Opin. Chem. Biol.*, **5**, 51–56.
8. Arai M and Kuwajima K (2000) 'Role of molten globule state in protein folding', *Adv. Protein Chem.*, **53**, 209–282.
9. Chasse G A, Rodriguez A M, Mak M L, Deretey E, Perczel A, Sosa C P, Enriz R D and Csizmadia I G (2001) 'Pepide and protein folding', *J. Mol. Struct. – Theochem.*, **537**, 310–361.
10. Clore G and Gronenborn A (1998) 'NMR structure determination of proteins and protein complexes larger than 20 kda', *Curr. Opin. Chem. Biol.*, **2**, 564–570.
11. Harris J and Craik C (1998) 'Engineering enzyme specificity', *Curr. Opin. Chem. Biol.*, **2**, 127–132.
12. Merry T (1999) 'Current techniques in protein glycosylation analysis – a guide to their application', *Acta Biochem. Polon.*, **46**, 303–314.
13. Fagain C (1995) 'Understanding and increasing protein stability', *Biochim. Biophys. Acta*, **1252**, 1–14.
14. Parodi A (2000) 'Protein glycosylation and the affects of glycosylation on protein structure and activity', *Chembiochem*, **1**, 215–246.
15. Radford S (2000) 'Protein folding: progress made and promises ahead', *Trends Biochem. Sci.*, **25**, 611–618.

16. Robertson A and Murphy K (1997) 'Protein structure and the energetics of protein stability', *Chem. Rev.*, **97**, 1251–1267.
17. Schulein M (2000) 'Protein engineering of cellulases', *Biochim. Biophys. Acta – Protein. Struct. Mol. Enzymol.*, **1543**, 239–252.
18. Wilmouth R C, Clifton I J and Neutze R (2000) 'Recent success in time-resolved protein crystallography', *Natural Products Rep.*, **17**, 527–533.

1.11.3 Web sites

19. ExPASy (analysis of protein sequences and structures). *www.expasy.ch/*
20. Protein Data Bank (worldwide repository for the processing and distribution of 3-dimentional biological macromolecular structure data). *www.rcsb.org/pdb/*
21. Enzyme Nomenclature Database (repository of information relative to the nomenclature of enzymes primarily based on the recommendations of the Nomenclature Committee of the International Union of Biochemistry and Molecular Biology). *www.expasy.ch/enzyme/enzyme_details.*
22. SWISS-PROT (probably the most widely used protein database) on the site of European Bioinformatics Institure. *www.ebi.ac.uk/swissprot/*

2
Substrates and their structure

G. BUSCHLE-DILLER
Auburn University, USA

This chapter is divided into two major parts. The first part (2.1) is directed towards non-fibrous enzymatic substrates or, in other words, compounds that undergo reactions with enzymes. Included in this part are dyes, natural sizing and thickener materials, such as starches and cellulose derivatives, and so on. Also in this part is a description of compounds that might be present but are not true substrates for enzymes. Their presence, however, may interfere with enzymatic recognition or otherwise affect the reactivity of the enzymes. The second part (2.2) deals with textile fibers that function as substrates for enzymes.

2.1 Non-fibrous substrates and non-substrates

2.1.1 Colorants for textile materials

Colorants generally are divided into dyes and pigments. While dyes for textile materials have moderate to good solubility in the medium of application, allowing them to penetrate into the fiber, pigments are insoluble in the medium of use and have to be attached to the fiber surface by other means, such as binders. Polymeric dyes fall in-between these two categories. Dyes can be classified by chemical class or by application method. Details of dye structure, application and end-use performance are listed in the Colour Index (CI) (Colour Index, 1999). A practical guide for dyeing and for printing can be found in Rivlin (1992) and in Miles (1994), respectively. For more theoretical background, a good reference is Zollinger's Color Chemistry (Zollinger, 1991). Dyes for specific fibers are discussed in Lewis (1992) for wool and wool blends and in Shore (1990a, 1995) for cellulosic fibers.

Efforts have been made to decolorize and/or fragment the chromophore of colorants on textile materials for bleaching or fading purposes, such as indigo on denim with laccases (Campos *et al.*, 2001). Furthermore, research was performed on dyehouse effluent treatment of aqueous suspensions/

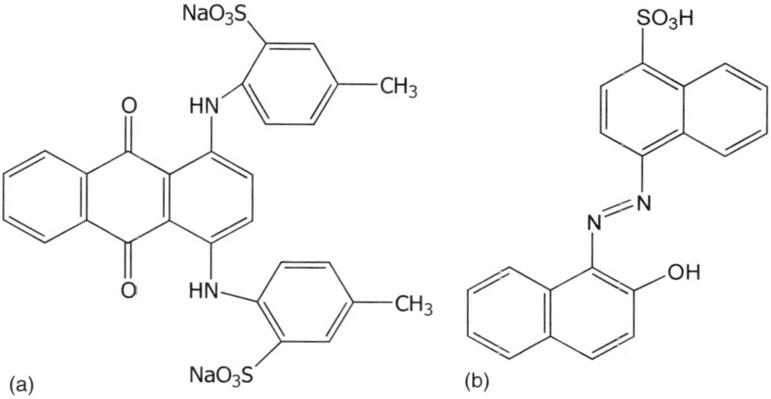

2.1 (a) CI Acid Green 25 and (b) CI Acid Red 88 as examples of acid dye structures.

solutions with peroxidases and laccases in the presence or absence of mediators (Call and Muecke, 1997; Tzanov *et al.*, 2000; Abadulla *et al.*, 2000).

In the following section, textile colorants as classified by their application method will be discussed briefly in alphabetical order of the dye class, followed by azoic and polymeric dyes as transition to pigments.

2.1.1.1 Acid dyes

Acid dyes are water-soluble dyes for polyamide fibers, wool and silk. The chromophore is often based on anthraquinone or one to several azo-groups and varies in molecular size (Fig. 2.1). Characteristic to all acid dyes are one or more sulfonic acid or carboxylic acid group(s), attached to the chromophore, which not only provide water solubility, but also form ionic bonds with the amino groups of the fiber polymer under acidic conditions. They show generally good washfastness and brilliant color shades.

1:1 and 1:2 metal complex dyes are a subcategory of acid dyes (Fig. 2.2). Their structural features allow for holding a transition metal ion via formation of a complex.

In 1:1 metal complex dyes, one dye molecule with two ortho-hydroxyl substitutions adjacent to the azo-group is coordinated with one metal ion, mostly chromium(III) or cobalt(III). When applied to the fiber, the three free coordinations of the metal ion interact with groups in the fiber, thus supporting dye–dye aggregation, the overall dye/fiber binding strength and ultimately the wash fastness. In 1:2 metal complex dyes, two dye molecules, which can be of the same or different type, form an octahedral arrangement with one transition metal ion in the center. In this case the metal ion is not available for additional bonds with the fiber since it is already fully coor-

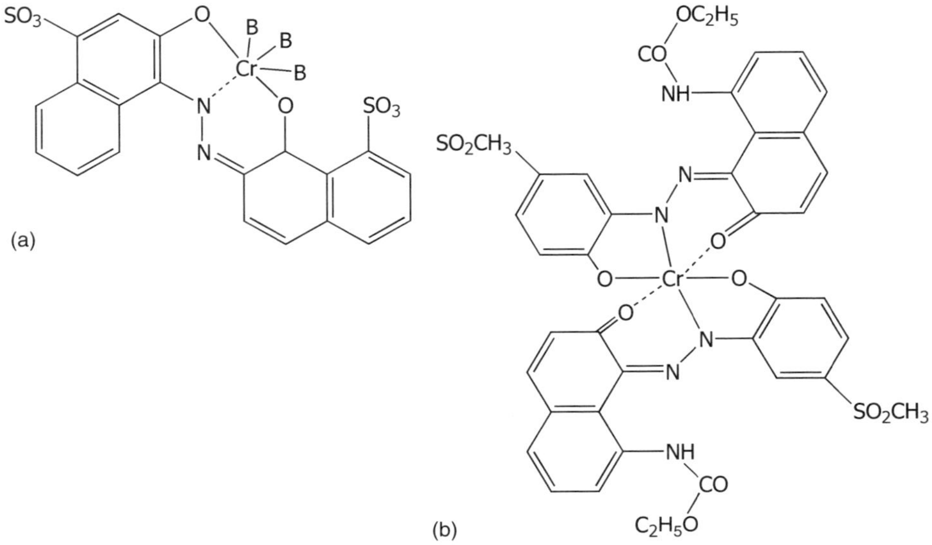

2.2 (a) CI Acid Blue 158 (B represents replaceable groups, such as water molecules) is an example of a 1:1 metal complex dye; (b) CI Acid Black 58 is an example of a 2:1 metal complex dye.

dinated. Dye/fiber binding in this case is improved through the dye molecular size and by hydrophobic interactions of these fairly large dye complexes with the fiber.

A similar subcategory comprises mordant dyes. Owing to environmental concerns as well as possible fiber damage, mordanting is no longer practiced commercially to any extent today. In mordant dyeing, a mordant solution is applied prior to acid dyeing to obtain various color shades. Popular mordants included salts of aluminum, iron and copper (Schweppe, 1993) and, dependent on the compound selected, a wide range of hues could be generated on wool, but also on linen and cotton. The most common mordant, however, was chromium applied in the form of sodium or potassium dichromate solution to wool in either a pre- or a post-treatment. The reduction of chromium(VI) to chromium(III) takes place in the fiber through cystine, by the way of hydrogen sulfide. Other groups in wool are possibly also involved in the reduction/oxidation process (Lewis, 1992).

2.1.1.2 Basic (cationic) dyes

Many naturally occurring dyes belong to the class of basic dyes. Examples are Crystal Violet (C.I. Basic Violet 3; Fig. 2.3a) and Malachite Green (C.I. Basic Green 4). The characteristic feature of basic dyes is a positively charged group within the molecule, either localized and fixed at a specific

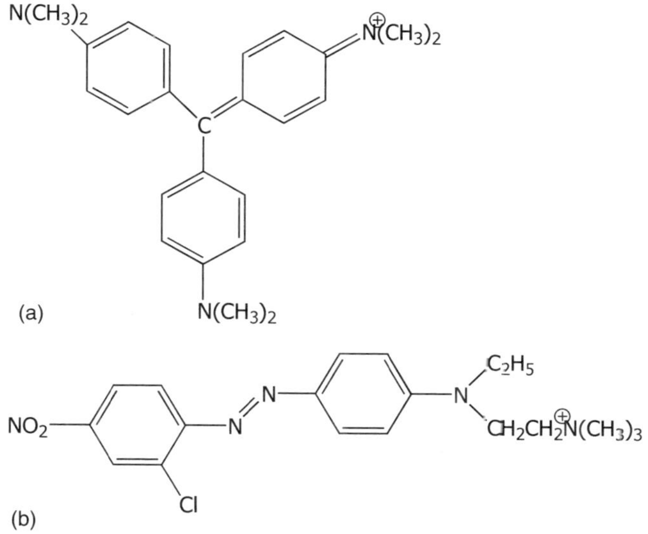

2.3 (a) Crystal Violet (CI Basic Violet 3) with delocalized positive charge and (b) CI Basic Red 18 with localized positive charge.

position (e.g. C.I. Basic Red 18; Fig. 2.3b) or delocalized over the entire structure (e.g. triarylmethane dyes). Historically used on silk, the major application of basic dyes today is focused on acrylic fibers and basic-dyeable polyester where strong ionic bonds with the acid groups incorporated in the fiber can be formed.

2.1.1.3 Direct dyes

Direct dyes are fairly high molecular weight, water-soluble, anionic dyes with substantivity for cotton and other cellulosic fibers. Direct dyeing is a simple exhaust process from neutral solution at boiling temperatures with the addition of an inorganic salt. Direct dyes are held inside the fiber through dye–dye aggregation, as well as by weak secondary forces formed by parallel alignment of dye molecules and cellulose chains. Owing to the size of their chromophores (Fig. 2.4), color shades are less brilliant and washfastness is limited. After treatments with cationic fixing agents, such as quaternary ammonium compounds, complex fatty acids and melamine derivatives, or crosslinking with amide–formaldehyde resins the wet performance of direct dyed goods can be improved.

2.1.1.4 Disperse dyes

Disperse dyes consist of small chromophores with strong auxochromes (see Fig. 2.5 for examples) and show very low water solubility, hence their name.

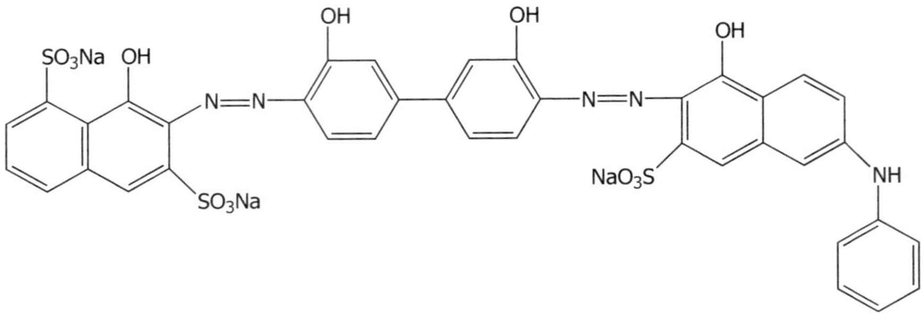

2.4 CI Direct Blue 98.

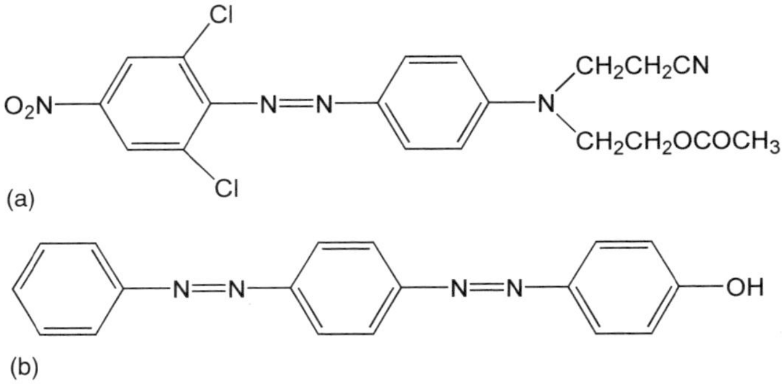

2.5 (a) CI Disperse Orange 30 and (b) CI Disperse Yellow 23.

They are the major dye class for polyester and other hydrophobic fibers, such as acetate, and to a lesser extent nylon. Several routes can be taken for the coloration with disperse dyes. If applied from an aqueous dispersion in an exhaust process, dyeing proceeds very slowly. Pressure and/or heat or the addition of so-called carriers are necessary measures to increase the dyeing rate. Carriers are small organic compounds that both support the dispersion properties of the dye as well as improve the accessibility of the fiber for the dye. The use of carriers is minimized as much as possible though, owing to their toxicity, odor and other undesirable properties.

An alternative route for application of disperse dyes is by a pad-dry-cure process (thermosol process). After padding the fairly concentrated dye dispersion onto the fabric, the remaining water is removed and dry heat close to the glass transition temperature of the fiber causes the sublimation of these dyes into the fiber structure.

2.1.1.5 Reactive dyes

The main characteristic of this water-soluble dye class is their ability to form a covalent bond with the fiber via a chemical reaction. Most of these dyes

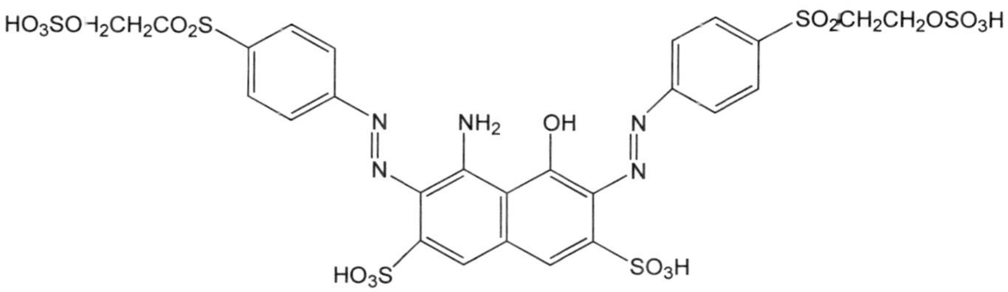

2.6 CI Reactive Black 5, bifunctional reactive dye.

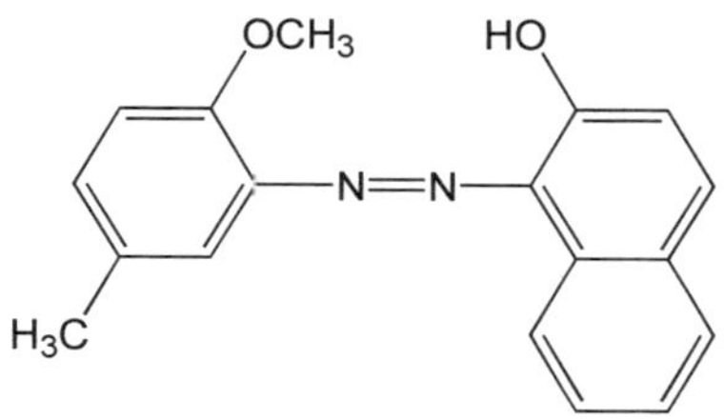

2.7 CI Solvent Red 17.

have been developed for cellulosics, but some special compounds also exist for the dyeing of wool (Lewis, 1992). In a reactive dye one or more reactive groups (same or different types) carrying a 'leaving' group are connected to the chromophore via a link (the different components of a reactive dye are shown in Fig. 2.6). The chromophore carries one or more water-solubilizing groups. The reaction with the fiber generally occurs by displacement of the leaving group (usually a halogeno substituent) at alkaline conditions via a nucleophilic substitution or elimination/addition mechanism depending on the type of reactive group. Under normal end-use conditions the stability of the formed covalent bond in cellulosics is good. Owing to the sensitivity to high pH, wool is dyed at neutral to slightly acidic conditions with especially developed reactive dyes. An alkaline aftertreatment with ammonia might be necessary.

2.1.1.6 Solvent dyes

Dyes in this class are water insoluble but can be dissolved in organic solvents, such as alcohols, ketones or hydrocarbons, for application. Their molecular weight is fairly low (Fig. 2.7) and their end-use properties (washfastness, etc.) vary considerably. Many solvent dyes find alternative uses in industries other than the textile industry.

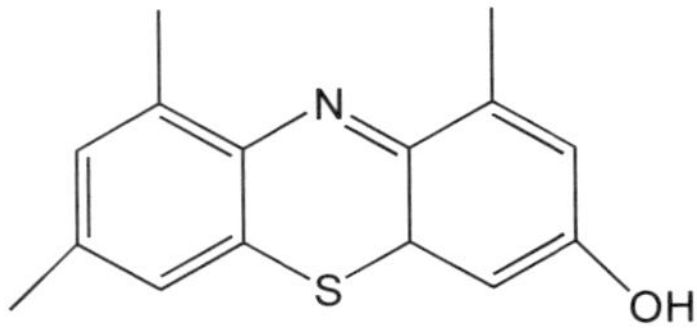

2.8 Example of a sulfur dye fragment.

2.1.1.7 Sulfur dyes

Sulfur dyes are water-insoluble polymeric dyes with sulfur as an integral part of the chromophore and sulfur in the form of polysulfide bridges within the dye structure, hence their name (see Fig. 2.8). The exact chemical structure and composition of most sulfur dyes is unknown. Under alkaline conditions internal S—S bridges can be split with a reducing agent, rendering the resulting dye fragments water soluble (*leuco* form of the dye) and with substantivity for cellulosic fibers:

$$\text{Chromophore-S-S-chromophore} + Na_2S \rightarrow {}^2\text{chromophore-}S^-Na^+$$

Once absorbed by the textile material, the dye fragments are reoxidized and combined to large network-like structures. Because of their insolubility and their molecular size they have good washfastness and thus are valuable in combined application with direct dyes, but their shades are dull and the color range obtainable is limited to browns, greens, blues and blacks.

Sulfur dyes are available in different forms on the market. Leuco sulfur dyes contain both the prereduced soluble *leuco* form of the dye and the reducing agent, such as sodium sulfide, and can be applied directly to the fiber. Solubilized sulfur dyes are thiosulfuric acid derivatives that can be converted to the alkali-soluble thiol form and exhausted onto the fiber. Condense sulfur dyes are S-alkyl- or S-arylthiosulfates which need sodium sulfide or polysulfide for reduction and solubilization.

2.1.1.8 Vat dyes

Application and manufacture of vat dyes have been known for centuries; indigo and Tyrian Purple (6,6′-dibromo indigo) are the most famous historic representatives of this dye class (Fig. 2.9). Similarly to sulfur dyes, vat dyes have to be made water soluble by reduction of the keto group in alkaline solution to the enolate form. Once in water-soluble form (leuco vat dye) they have sufficient substantivity for cellulosic fibers. Reoxidation with air or a mild oxidizing agent reforms the original water-insoluble dye molecule. Vat dyes show good washfastness and fairly brilliant colors. They can also be obtained in their prereduced salt form (sulfatoesters of vat dyes) for pH sensitive fiber materials (wool, blends) and easier application. However, the shelf life of these soluble vat dyes is limited.

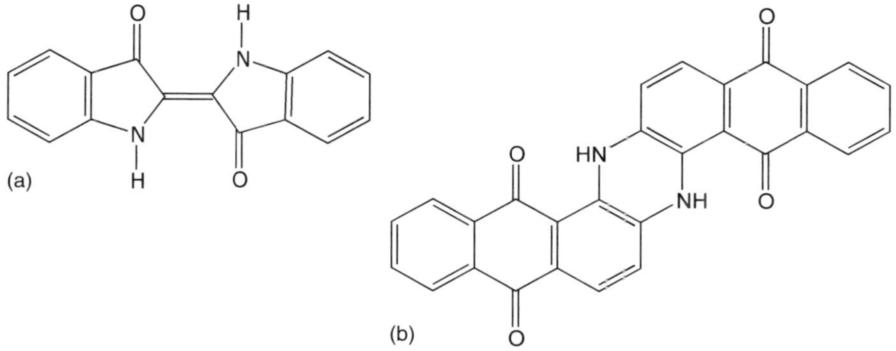

2.9 (a) CI Vat Blue 1 (indigo) and (b) CI Vat Blue 4.

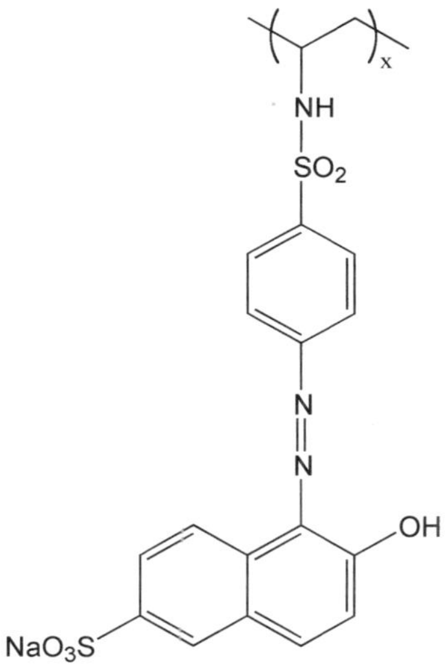

Poly S™-119

2.10 Structure of a commercial polymeric dye.

2.1.1.9 Polymeric dyes

In the case of polymeric dyes, the chromophore is attached to a polymer backbone, such as polyethylene glycol (Fig. 2.10). A review of polymeric dyes (i.e. polymers that contain chromophoric substituent groups or polymers which contain a chromophoric group as part of the polymeric backbone), emphasizing synthesis and characterization, can be found in Guthrie (1990).

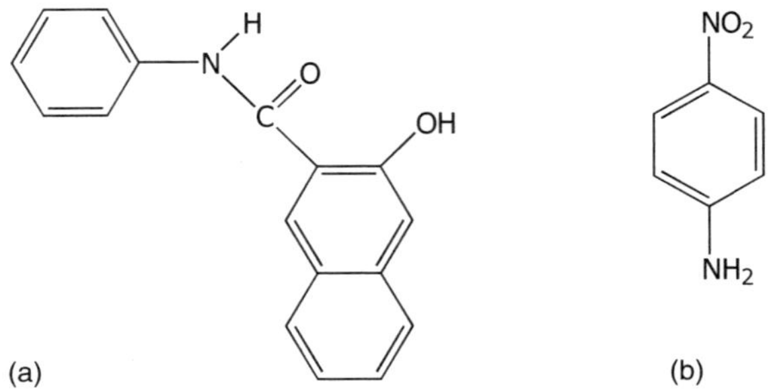

2.11 Structure of (a) CI Azoic Coupling Component 2 and (b) CI Azoic Diazo Component 37.

2.1.1.10 Azoic dyes

Azoic dyes are relatively small, mostly monoazo, compounds that are synthetized on the fiber by coupling a water-soluble diazo compound and a water-soluble coupling component (Fig. 2.11). Coupling components are substituted β-naphthols with affinity for cellulosic fibers which have been made water-soluble in alkaline solution (Colour Index terminology: Azoic Coupling Component). Diazo compounds (Azoic Diazo Component) are aromatic primary amines or stabilized diazonium compounds that are reacted to diazonium salts under acidic conditions before coupling in a separate bath. The naphthol is first applied to the textile material and, after removal of excess liquor, developed in diazonium salt solution. Although the application process is fairly complicated and time consuming, azoic dyes have some advantages over other dye classes. Color shades obtained are especially strong and brilliant in the red and orange area, but also in blue and black. The dye formed is water insoluble with high wash fastness. However, crock fastness is limited, the reason being that the dye formed on the surface resembles a pigment, lacking the binder to fix it properly. Surface dye particles are thus easily removed upon mechanical action. Today azoic dyes are mainly used for coloring cellulosic materials, but can also be applied to polyester and triacetate.

2.1.1.11 Pigments

Pigments are organic or inorganic colorants (see Fig. 2.12 for an example of an organic pigment) that are insoluble in the medium of application and thus remain in the particle state. If used on textile materials they are fixed to the surface with the help of binders or in the form of coatings or mixed into the spinning dope of synthetic fibers. Suitable adhesives often consist of resins that can be heat set, holding the pigments in place. Fastness to washing, dry cleaning and especially crocking (rubbing) of

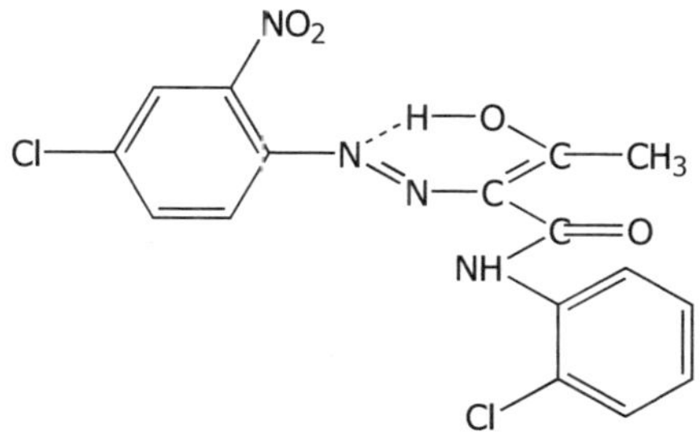

2.12 Cl Pigment Yellow 3.

pigment prints are usually determined by the wear-and-tear properties of the binder used.

2.1.2 Natural auxiliaries for textile processing

The purpose of auxiliaries is to facilitate a textile process and/or increase its efficiency. They serve as sizing materials, lubricants, wetting agents, emulsifiers, agents accelerating or decelerating the dyeing rate, thickeners, binders, etc. often with considerable overlap in the functions and abilities of a specific chemical. Compounds used encompass many different chemical classes, some of which are affected by enzymes and thus can be regarded as substrates, and some of which remain unaffected. Owing to environmental and economical concerns, auxiliaries are used as sparingly as possible. Once the respective process is terminated they are to be removed completely from the treated material; however, traces could still be present and interfere negatively with subsequent processing steps.

2.1.2.1 Natural sizing compounds, coating materials and thickeners

Starch

Sizing compounds and lubricants are applied to yarns before fabric formation to protect the integrity of the yarns. While increasingly faster weaving processes demand more enduring sizes, such as acrylic-based compounds, natural sizes that can be decomposed by enzymes are still on the market. Such compounds comprise starch and starch derivatives, as well as soluble cellulose derivatives, with waxes often admixed. Desizing with amylases is one of the oldest enzymatic processes used in the textile industry. A comprehensive description of the process can be found in Uhlig (1998).

Starch has also been very useful as a thickener in printing pastes and as a component of adhesives. In printing processes, starches are applied to

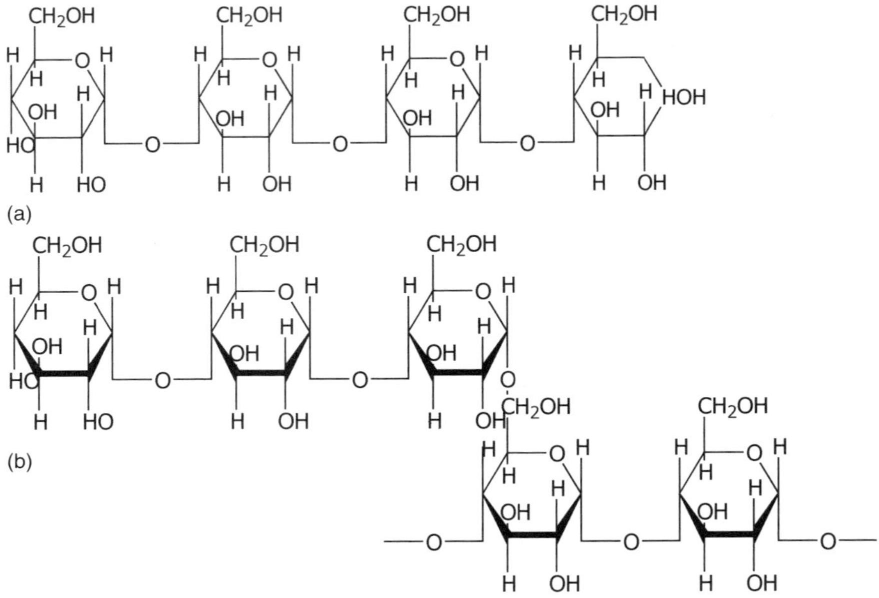

2.13 (a) Amylose and (b) amylopectin, the two components of starch.

guarantee a defined design and to avoid spreading of the printing paste. In the paper industry, starches increase sheet strength and, as coatings, improve the writing and printing properties of high quality paper.

Native starch contains two components, amylose and amylopectin (see Fig. 2.13), bound together by hydrogen bonding. Degrees of polymerization and cross-linking vary, yielding a large variety of starches with differing characteristics, including swelling and gelatinization properties. Amylose is a polymer of linear unbranched α-1,4-glucan, while amylopectin, also a α-1,4-glucan, is additionally highly branched at the C-6-position. The ratio of amylose and amylopectin depends on the source and processing of the respective starch.

Enzymes capable of hydrolyzing starch include α- and β-amylase, amyloglucosidase (glucoamylase) and isoamylase (Uhlig, 1998). Both α- and β-amylases attack the α-1,4-linkage, but are unable to break the 1,6-branched linkages. They predominantly produce maltose and dextrins as end products. Glucoamylase liberates glucose from non-reducing ends at α-1,4 and 1,6-linkages and generates glucose. Isoamylase is a debranching enzyme, producing mainly maltose.

Modified starches

The hydroxyl groups in starch can be functionalized to form acetates, ethers or esters to various degrees of substitution. Such modifications have an impact on the gelation and swelling behavior and are useful for printing applications as well as in the food and pharmaceutical industry. Among others, hydroxyethyl-, methyl- and carboxymethyl-starches are important as thickeners. Corresponding water-soluble cellulose derivatives can also be produced for application in thickener formulations (see below).

Gums

Dry roasting of starches at 135–190°C yields British gums. Such starch derivatives are crosslinked in the 1,6 position with decreased degree of polymerization (DP). Other gums that are important for thickening purposes are alginate based (from seaweed) or obtained from other natural polysaccharides (gum arabic, xanthate gum, gum tragacanth, etc.; Miles, 1994). Some of these compounds can also economically be grown by microorganisms. Dextrins (see below) are produced by pyroconversion with very small amounts of acid in the form of a fine spray. Gums and dextrins play an important role as adhesives and binders in the coating industry and as encapsulating compounds.

Xanthan

Xanthan gum consists of a β-1,4 linked glucan main chain with a negatively charged trisaccharide (two mannose units and one glucuronic acid residue) side chain at alternating C3 atoms. Because of these substitutions, xanthan is a highly charged polyanion, giving it a rigid-rod structure in solution as a result of repulsion effects. Considering its structure and its high molecular weight, it is not surprising that its solution is highly viscous. Low concentrations are sufficient to prepare an excellent thickening agent for printing (Miles, 1994) and other purposes.

Alginates

Alginates are polyanionic block copolymers of high molecular weight of α-D-guluronic acid and β-L-mannuronic acid (De Baets *et al.*, 2002). Commercial alginates are obtained from seaweed. With the help of enzymes or by chemical means the ratio of guluronic acid and mannuronic acid can be modified and thus their gelation properties tailored to specific applications. The high viscosity of their solutions and their chemical inertness towards dyes makes them valuable textile printing thickeners.

Dextrin and cyclodextrin

Dextrins are α-1,6-D-glucopyranosyl polymers, branched through (1,2), (1,3) or (1,4) linkages, while cyclodextrins are cyclic molecules formed by

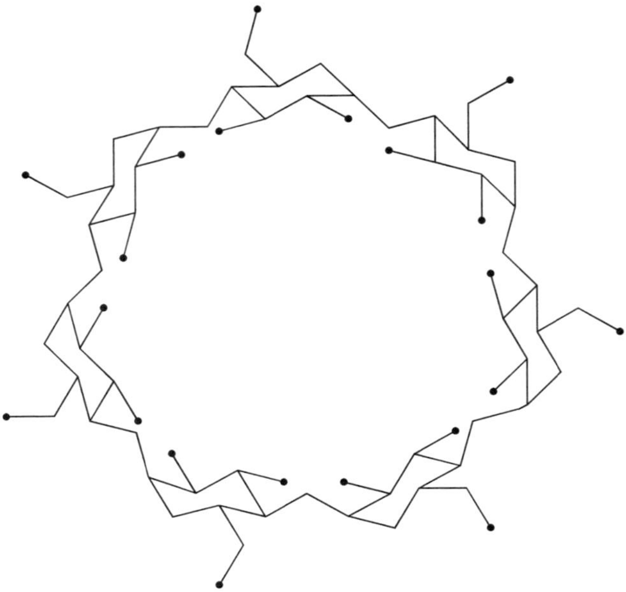

2.14 Structure of cyclodextrin, made from seven glucopyranose residues.

six, seven, eight or, less commonly, nine α-D-(1,4)-linked glucopyranose residues (see Fig. 2.14). Cyclodextrins form water-soluble ring structures with numerous hydroxyl groups at the outside. The interior cavity on the other hand is fairly hydrophobic and capable of holding small molecules. Besides sequestering small molecules, cyclodextrins can serve as amylase inhibitors.

Cellulose derivatives

Water-soluble cellulose ethers (Fig. 2.15) are versatile auxiliaries owing to their availability and low toxicity. Carboxymethyl cellulose, and methyl (MC) and hydroxyethyl celluloses (HEC) are commercially important thickeners, film formers, adhesives and water-retaining agents for the textile, pulp and paper industries (Heinze, 1998). Organic and inorganic cellulose esters are used in coatings, thermoplastic films and resins (e.g. cellulose acetobutyrate) and for textile finishing (e.g. cellulose phosphate for flame retardancy for cotton). Cellulose acetate can also be produced in textile fiber form and is discussed in Section 2.2.1.

Cellulases are suitable enzymes for the decomposition of cellulose derivatives on condition that the derivatization of the cellulose backbone does

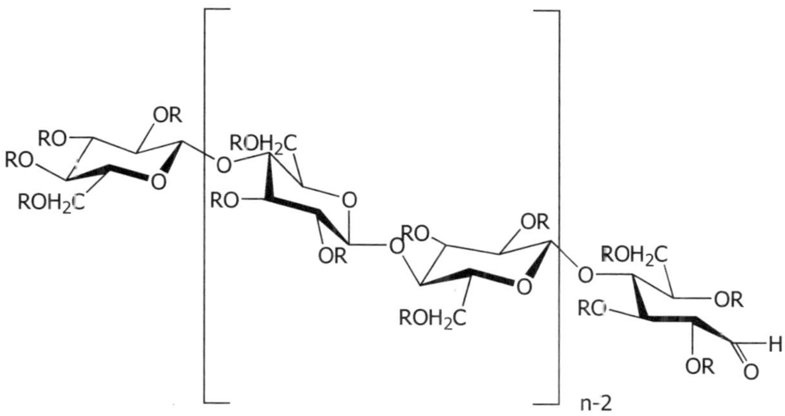

Cellulose derivatives (some R = H, some R = ester-, ether-, etc. forming groups)

2.15 Modified celluloses (some rest groups R=H, some or all R=CH_3, C_2H_5, $COCH_3$, etc.).

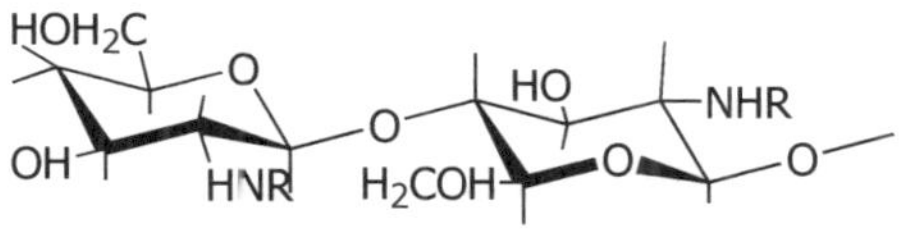

2.16 Chitosan (R=H) and chitin (R = acetate).

not have a significant impact on enzyme recognition (Philipp and Stscherbina, 1992; Glasser *et al.*, 1994). Whole cellulases, composed of cellobiohydrolases, endocellulases and β-glucosidases, can be applied either individually or in combination with enzymes that focus on other compounds in the thickener or sizing material, such as amylases (starch degradation) or lipases (fat hydrolysis). In enzyme mixtures the requirements for active pH and temperature of all components must be considered for the system to be effective.

Chitosan

Chitosan (copolymer of *N*-acetylglucosamine and glucosamine, Fig. 2.16; De Baets *et al.*, 2002) is produced from chitin (β-1,4-linked *N*-acetylglucosamine) by deprotonization, demineralization and partial deacetylation to a product soluble in 1% acetic acid. Chitosan can be used as a viscosity controlling compound in mixtures with other swelling agents. Further, as a polycation, chitosan binds to polyanions such as anionic dyes, and can improve dyeability if applied as a coating or film forming agent. It is possible to manipulate film stabilities by crosslinking. Chitosan oligomers of DP >30 have antimicrobial properties and thus are useful as wound dressings as well as in finishing of apparel and household goods (Kumar, 2000). As a paper coating or mixed into pulp, paper sheet strength can be improved.

Chitinases (poly-β-glucosaminidases) hydrolyze 1,4-β-linkages of *N*-acetyl-D-glucosamine polymers of chitin. They are produced by microorganisms and by plants, such as soybeans and tomatoes, and consist of three types of enzymes, endo- and exochitinases and chitobiase. These enzymes catalyze the breakdown of cell walls of organisms with glucosamine polymeric structures. Chitosanases cause endohydrolysis of 1,4-β-linkages only in polymers with 30–60% acetylation (Fukamizo, 2000).

2.1.2.2 Spin finishes, natural lubricants, oils, waxes

Spin finishes fulfill either the task of increasing fiber friction in the form of cohesive agents or of decreasing fiber friction and softening. Natural sources of lubricants are fatty acids obtained by saponification of fats and waxes, such as coconut, cotton seed, peanut, corn or palm oil, butter fat, lard or beef tallow. Commercial products for fiber lubrication usually contain mixtures of various natural and/or synthetic compounds (Slate, 1998). Oils and fats can be hydrolyzed by superheated pressurized steam. Alternatively, enzymatic hydrolysis with lipases partially or fully achieves the degradation of fats and oils to free fatty acids and glycerol (Uhlig, 1998).

2.1.3 Synthetic auxiliaries for dyeing and finishing

2.1.3.1 General

The function of auxiliaries in Section 2.1.2 including compounds that are affected or degraded by enzymes was discussed. Surface active substances, salts, oxidizing and reducing agents, and acids and bases also belong to the category of auxiliaries; however, they are not considered enzymatic substrates although they might influence the effectiveness or mode of action of enzymes.

2.1.3.2 Electrolytic compounds and pH control substances

For all textile processing steps, water quality and softness, pH and electrolyte content are important considerations. Water hardness is caused by calcium and magnesium sulfates, chlorides (permanent hardness) and carbonates (temporary hardness). These salts not only contribute to deposits on equipment, but also interfere with preparation, dyeing and finishing. Various techniques are available to soften water on an industrial level, most commonly via ion exchange. A more direct and more expensive approach is the addition of sequestering agents to process water where water softness is crucial. Sequestering agents have functionalities that allow complexing (chelating) of metal ions. Examples for such chelators are polycarboxylic acids (e.g. oxalic acid), aminopolycarboxylic acids (EDTA, ethylenediaminetetra-acetic acid), sodium polyphosphates (sodium hexametaphosphate, Calgon®), and others.

Besides water softness, the pH of the treatment bath in preparation, dyeing and finishing plays an important role owing to the sensitivity of the textile material to acid or basic conditions on the one hand and the reactivity of dyes and finishing compounds on the other hand. Many processes even require the stabilization of the pH with the help of a buffer system. Buffer systems consist of an acid and the corresponding salt, for example, acetic acid and sodium acetate for pH 4–5, or a base and the corresponding salt. Buffer systems for any type of pH range can be found in general laboratory reference books (e.g. Shugar and Ballinger, 1990). Common acids are inorganic acids (e.g. sulfuric acid, hydrochloric acid, pH below 2), organic acids (acetic acid, citric acid, pH 4.5–3.5), acidic inorganic salts (ammonium sulfate, etc. for pH 6.5–5.5) and mixtures thereof. Alkaline pH ranges are adjusted with common bases (sodium or potassium hydroxide, pH 11 or higher) or basic salts (e.g. carbonates, borax). Great care has to be exercised to make sure that adequate rinsing takes place after each treatment step and that sufficient time is allowed for internal exchange processes within the fiber. Because of adsorption processes, the release of acids, bases or salts can be fairly slow, and the pH of the rinse bath might not represent the realistic pH situation inside the fiber pores.

A large number of dyeing processes afford the addition of common salts, such as sodium chloride or sodium sulfate to enhance dye adsorption and fixation. The amount of these salts can be quite considerable, often 10% of the weight of fiber or more. The function of these salts is first to help alleviate negative fiber surface charges, thus reducing repulsion between negatively charged dye molecules and the fiber wall. Second, they support the aggregation of dye molecules inside the fiber pores by making the dye less ionic (for example, in direct dyeing of cellulosics). In some cases, an example being acid dyeing of wool, salts act as retarders. The smaller salt ions temporarily take the place of the dye at the fiber dye-site. They are then replaced by the larger and slower moving dye molecules, yielding much more uniform dyeing results. Thorough rinsing has to follow any dyeing process, not only to remove unbound and loosely attached surface dye, but also to eliminate any auxiliaries. Even minute amounts of remaining salts can show up as white deposits on dyed goods.

2.1.4 Compounds with whitening effect

High levels of whiteness are desirable for textile materials as well as fundamental for reproducible color shades. Thus, whitening is usually carried out prior to dyeing and finishing. Most commonly this is achieved by the use of oxidizing agents that destroy chromophoric substances. Additionally, fluorescent brightening agents are added that mask yellowing compounds. Natural cellulosics are in most need of bleaching. Most synthetics are already fairly white; however, if necessary, fluorescent brighteners can be included in the spinning dope. Wool and silk are not routinely bleached.

Common bleaching agents include hydrogen peroxide and chlorine-containing compounds (Trotman, 1984; Shore, 1990b). Hydrogen peroxide is preferred over other bleaching agents as it decomposes into oxygen and water without impact on the environment. Bleaching is performed at pH 10.5–11 at boiling temperatures in the presence of stabilizers, sequestrants to control water softness and metal content, and surfactants with detergency. Stabilizers often consist of polysilicates, acrylates or magnesium salts.

The mechanism of bleaching most likely follows a radical route (Zeronian and Inglesby, 1995). In the presence of metal ions that act as peroxide activators, fiber damage is possible as radicals can attack the fiber polymer instead of the chromophore of the colorant. Small amounts of hydrogen peroxide that might still be held back by the fiber after bleaching have to be removed to avoid interference later on with dyes or finishes. Besides by chemical means, this step can also be performed enzymatically with catalase, an oxidoreductase that catalyzes the breakdown of hydrogen peroxide to water (Tzanov *et al.*, 2002).

The bleaching effect of sodium chlorite strongly depends on the pH value. The reaction occurs most rapidly at low pH and higher temperatures. In commercial operations bleaching is performed at pH 3.5–6 for cotton and temperatures around 80°C. Toxic chlorine dioxide is produced at lower pH; above pH 9 the bleaching effect is insignificant.

Sodium hypochlorite bleaches at pH values above 11 in a buffered system. The active chlorine content should be determined before use. Severe oxidative fiber damage can be expected if the pH falls below 9 with formation of hydrochloric acid, which will reduce the pH despite the buffer system. Further, hypochlorite decomposes upon storage or exposure to light, and an antichlor after-treatment with reducing agents following treatment with chlorite and hypochlorite might be necessary to remove chlorine traces from the fiber. Owing to the possible encounter of significant problems with hypochlorite and chlorite, hydrogen peroxide has advanced to the favored bleaching agent in commercial wet-processing operations nowadays. Still, chlorine-containing agents are sometimes applied to bleach bast fibers, such as flax or jute.

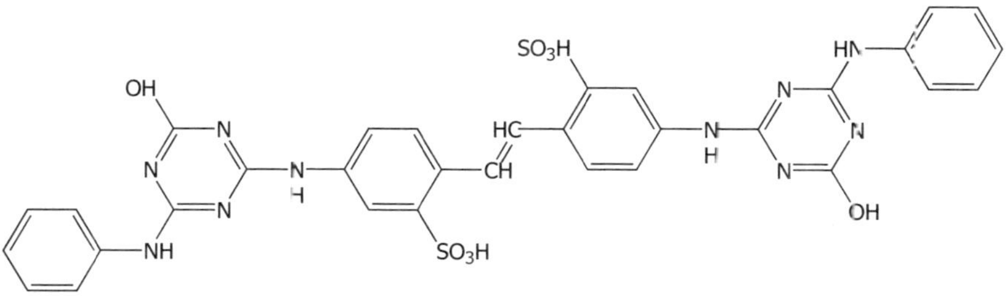

2.17 CI Fluorescent Brightener 32 for cotton.

Fluorescent brighteners (optical whiteners) for cellulosics are essentially colorless direct dyes (Fig. 2.17) and colorless acid dyes are used for wool. These compounds mask yellowness in fibers by absorbing light in the ultraviolet (UV)-range and emitting it in the violet or blue region of the visible light, thus compensating for minor yellowness of the textile substrate. Fluorescent brighteners and bleaches are often applied together. Many of these 'dyes' are not ultimately washfast on natural fibers and are reapplied with the laundry detergent at each domestic wash. Fluorescent brighteners for synthetic fibers can be incorporated in the spinning dope and, in this case, are permanent. Photoyellowing of textile material can occur faster if fluorescent brighteners are present as a consequence of the breakdown or photomodification of its chromophoric system. These light-induced processes can then turn the fluorescent brightener into a regular dye.

2.1.5 Compounds intended to affect interfacial properties

2.1.5.1 Surfactants

Surface-affecting substances (surfactants) are a very important group of textile auxiliaries. They find use as wetting agents, softeners, detergents, emulsifiers and defoaming agents, to name just a few applications. Commercial products rarely contain a pure compound, but rather mixtures of a range of surfactants to tailor their properties to the tasks in demand (Flick, 1993).

Surfactants generally consist of a hydrophilic part, providing water solubility, and a hydrophobic part, creating a link to non-aqueous media. Based on the nature of the hydrophilic portion of the molecule they are classified as follows (Broze, 1999):

- Anionic surfactants: negatively charged groups (e.g. sulfates, carboxylates, phosphates or sulfonates) are associated with the hydrophobic part of the molecule. These surfactants are important wetting agents and detergents. Owing to the fact that many substrates are also negatively charged, anionic surfactants do not firmly adhere to such surfaces and impede redeposition of soil.

- Non-ionic surfactants: polar but without actual charge, solubilization properties in non-ionic surfactants are usually provided by incorporation of ethoxy units into their structure (alcohols, ethers, esters, etc.). Non-ionic surfactants can be mixed with any other group of surfactants and are fairly insensitive to water hardness. Most commonly, they are blended with anionic surfactants for increased detergency or used alone as emulsifiers.
- Cationic surfactants: these surfactants carry a positively charged group, commonly a quaternary ammonium group, associated with the hydrophobic portion of the molecule. Often, these compounds are additionally ethoxylated. Being positively charged, cationic surfactants adsorb more firmly to negatively charged substrates. Their major application is in softeners and emulsifiers.
- Amphoteric (zwitterionic) surfactants: these surfactants contain both anionic and cationic groups in their structure and thus behave as cationic or anionic compounds dependent on the respective pH. Common structures are betaines, amino acid derivatives and imidazoline derivatives. Their isoelectric point does not necessarily lie at pH 7. Although they seem to have a great application potential, they are the least important commercially.

The hydrophobic portion in all four types of surfactant consists of fairly long-chained linear saturated or unsaturated alkanes, derived from fats or oils. The chain length lies between 8 to 18 carbon atoms (e.g. stearate, palmitate, oleate, linoleate). Aromatic moieties and/or alkyl-substituted groups are also common.

Surfactants added in increasing amounts to water orient themselves at the interface of water/air with their hydrophilic parts towards the water and the hydrophobic parts pointing into the air. At a specific concentration (critical micelle concentration), when the entire water surface is covered by surfactant molecules, more or less ordered aggregations of surfactant molecules form in the bulk of the solution (micelles). In a micelle the polar hydrophilic parts of the surfactant molecules are oriented towards the water, the hydrophobic parts towards the interior of the micelle (in oil instead of water, their orientation is reversed). The hydrophobic center of the micelle can thus interact with hydrophobic compounds of the system, such as insoluble dyes, finishes, oils, etc., fulfilling solubilizing, emulsifying and dispersing tasks (Datyner, 1993).

Enzyme reactions on textile materials have been performed in the presence of the various types of surfactants and their effect studied. The reports, however, often provide controversial results (Helle *et al.*, 1993; Kaya *et al.*, 1995; Ueda *et al.*, 1994).

2.1.5.2 *Foam control substances*

Foam control is a very important issue for various textile processes, such as scouring, dyeing and printing, and wet processing with the goal of economic, low water pick-up. For a foaming system to be effective both foam and anti-foam agents are necessary to control the liquid drainage rate from the film walls. The interfacial tension between the foaming and defoaming compound needs to be manipulated. Anionic and non-polar surfactants can function as both types; however, fats, waxes, fatty acids and oils, long-chaine alcohols and polyglycols, polyalkylsiloxanes and their block copolymers with poly(oxyethylene) are more efficient as defoamers.

Foam application with controlled foam stabilization and collapse can be created with a blend of anionic and non-ionic surfactants and foam stabilizers, such as poly(vinyl alcohol), poly(acrylic acids), polysaccharides and cellulose derivatives.

Foam breakers are compounds that destroy foam, while foam inhibitors are made to prevent foam from being formed. Foam breakers quickly drain liquid and drastically reduce the surface tension at interfaces. They often consist of metal carboxylates in oil dispersion. Common formulations for effective foam inhibitors are water-soluble silicone glycol chemicals (see below), silica dispersed in water, or fluorinated alcohols and acids. Such compounds replace the elastic surface film by a more brittle film, so that the increase in surface tension caused by expansion is counterbalanced (Slate, 1998).

2.1.5.3 *Softening agents*

Besides cationic surfactants, often mixed with non-ionic surfactants, polysiloxanes are frequently used for fabric softening. Siloxanes are usually non-durable, but can be made durable by modification and addition of functional groups, followed by crosslinking. Other permanent types include reactive N-methylol derivatives of fatty acids and chlorotriazines, similar to reactive groups in fiber-reactive dyes. Some of these softeners are commonly applied together with easy-care finishes. Finishing compounds that render the textile material less hydrophilic by either coating the fiber surface and/or crosslinking and thus closing up the amorphous areas can present a barrier for enzymatic access.

2.1.6 Silicones and fluorochemicals

Silicones (Fig. 2.18) are compounds that can act as defoamers, soil repellants or lubricants. Fluorochemicals are especially useful as water- and stain-repellants. Both chemical classes encompass a wide range of compounds and are valuable for various purposes. Depending on the chemical

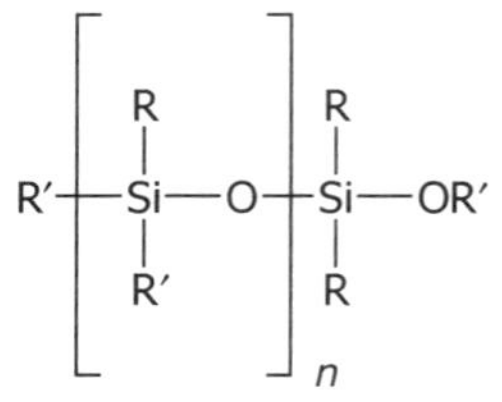

Silicone (R, R' = H, OH, CH_3, etc)

2.18 Silicone backbone.

composition, their viscosity and hydrophobicity differ. Most common are poly(dimethyl siloxane), poly(dimethylmethyl:phenyl siloxane) and poly(glycol/silicone copolymers). Such compounds are hydrophobic and add water repellency to cotton. However, they can increase the propensity for soiling. Modification with fluoroalkoxyalkyl groups gives compounds that yield water-, oil- and stain-resistant properties on textile materials. Besides modified silicone-based water- and stain-repellent compounds, effective fluorinated carbamates, fluorocarbon urethanes, polyfluorourea resins as well as fluoroalkyls combined with phosphates have been developed for cotton, wool and carpet fibers (Slate, 1998). Many of these compounds are proprietary.

2.1.7 Synthetic sizes and thickeners

A large group of synthetic sizes and thickening agents are acrylic-based polymers, either linear or crosslinked in structure. If used for sizing, they often remain on the fabric to add to the hand properties and softness of the textile material. Besides, complete desizing is often problematic (Lewin and Sello, 1983a; Lewin and Pearce, 1998).

Poly(vinyl alcohol) used as a synthetic size or thickener has the advantage of being easily recyclable and reusable as it can be removed by dissolving in hot water (Reife and Freeman, 1996). Polyacrylates swell in hot water and need sufficient mechanical impact to be completely removed from the fabric. Polyester-based sizes are broken down by hot alkaline solutions; however, insoluble oligomers may remain on the fabric (Shore, 1990b).

Copolymers of methyl methacrylate are soluble in organic solvents. Their application and removal occurs in non-aqueous media. If completed in a closed system, they are valued as environmentally benign; however, the machinery necessary needs modification from standard equipment to accommodate the process with solvents other than water.

2.1.8 Crosslinking resins

A large number of cellulosic fabrics, especially cotton and cotton in blends with polyester, are finished with easy-care or wrinkle-resistant finishes. Compounds used for this process form crosslinked networks involving the cellulosic hydroxyl functional groups on the fiber surface as well as in the accessible fiber interior, thus providing dimensional stability by fixing the structure in a specific state. Coloration of the textile material has to be performed prior to crosslinking because otherwise the amorphous areas become partially or completely inaccessible. If carried out on dyed material, crosslinking leads to improved wet fastness, locking the dye molecules in place.

Crosslinking resins are usually applied in monomeric or prepolymerized form together with a catalyst, and dried and cured for short times at fairly high temperatures (Vigo, 1994; Lewin and Sello, 1983b). As a general rule, the length of the crosslinks is determined by the moisture content and thus by the amount of swelling of the fiber. Therefore, in wet fibers the crosslinks are the longest, which may create some slack in the dried fiber. Wrinkle resistance is moderate to good and loss in tensile strength is limited. Crosslinking of dry fibers, on the other hand, yields short bridges, provides excellent wrinkle recovery and fairly high tensile strength losses. Hand builders are often added to improve the harsh feel of these finished goods.

The selection of compounds explored for this process is very large and is documented in thousands of publications and patents. A major group of crosslinking resins is based on urea and melamine–formaldehyde precondensates. Examples of common resins are dimethyloldihydroxyethylene urea (DMDHEU, Fig. 2.19), dimethylolethylene urea (DMEU) and dimethylolpropylene urea (DMPU). Newer compounds include polyfunctional carbamates, 4-alkoxypropylene ureas and *N*-methylolacrylamide derivatives (Vigo, 1994). Catalysts for these finishes are most often inorganic acids or salts, which can cause a drop in DP of the fiber polymer owing to the sensitivity of cellulosics to acidic conditions. Formaldehyde-free compounds include multifunctional carboxylic acids, such as 1,2,3,4-butanetetracarboxylic acid (BTCA, Fig. 2.19), citric acid and maleic acid (Raheel, 1998). Their finishing effect is somewhat less in most cases.

2.1.9 Flame-retardant finishes

Flammability of textile materials has always been a major problem and numerous attempts have been made to develop effective finishes to improve flame retardancy for all types of textile fibers (Lewin and Pearce, 1998). The approach taken involves delaying ignition, reducing the amount of flammable gases during a fire, increasing the amount of charred mater-

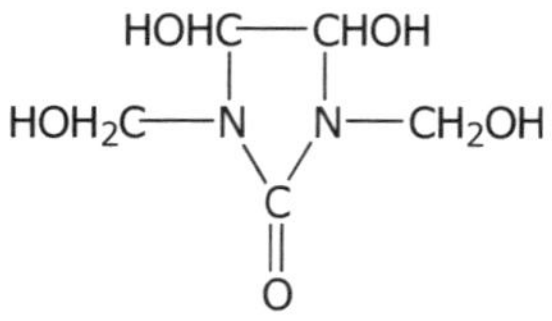

DMDHEU

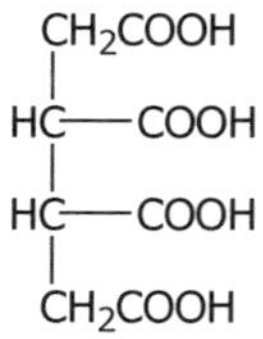

BTCA

2.19 Structures of DMDHEU and BTCA as examples of formaldehyde-containing and formaldehyde-free crosslinking agents, respectively.

ial formed, and enhancing the capability of a material to withdraw from the source of combustion (thermoplastic fibers). Depending on the fiber type, flame-retardant compounds can be applied as topical finishes, grafted onto the fiber, copolymerized during fiber formation, or incorporated into the spinning dope of synthetic fibers. The durability and effectiveness of these finishes vary. Commercially available compounds are based on a variety of inorganic metal salts (e.g. antimony, titanium, zirconium), on boric acid and its salts, phosphoric acid and its salts, organophosphates and halogen-containing compounds. Combinations of different compounds are found to have a synergistic effect (Lin and Zheng, 2002).

2.2 Textile fibers as substrates for enzymes

Classic textile substrates for enzymes are natural fibers, but a few synthetic fibers have also been subjected to enzymatic treatments. The major goal of most enzyme treatments is the modification of the fiber surface to enhance the hand and appearance of the fiber. Examples for such fiber modifications are the treatment of cotton denim with cellulases to achieve a soft jeans material with a washed-and-worn look, or biopolishing of cotton to brighten the colors and enhance the comfort. Other goals are to remove undesirable byproducts, such as pectins or fats, from the unscoured fiber or to soften woody material during retting with the help of suitable enzymes. The use of enzymes has also been explored in connection with shrink-proofing of wool and degumming of silk. In either case the enzymes of choice belong to the class of hydrolases and usually consist of multicomponent systems with a synergistic mode of action rather than of individual enzymes. Enzymatic processes offer major advantages over conventional treatments, including savings in chemicals and energy, and less or no impact on the environment. Additionally, if carefully controlled, they do not cause any fiber damage.

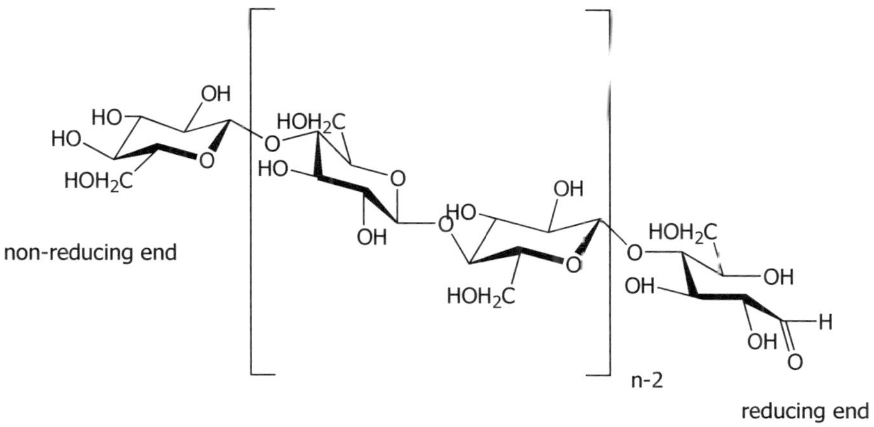

2.20 Structure of cellulose.

2.2.1 Cellulosic fibers

Cellulose is the most abundant renewable polymer today. In fairly pure form, cellulose occurs in the seed hairs of cotton plants, and is less pure in grasses and other plant material. Most cellulose, however, is found in the cell walls of woody plants together with lignin, hemicelluloses and other compounds as byproducts.

Cellulose is a linear 1,4-linked β-D-glucan homopolymer (Fig. 2.20) and constitutes the major component of higher plant cell walls. The monomeric unit is represented by cellobiose. The DP varies strongly dependent on the cellulose source and processing stage of the cellulosic material. Three free hydroxyl groups in C2, C3 and C6 per anhydroglucose unit (AGU) are available for formation of strong inter- and intramolecular hydrogen bonds as well as bonds/interactions with introduced compounds such as dyes and finishing agents. Further, all three or part of the primary and secondary hydroxyl groups can be chemically transformed into cellulose derivatives.

For textile materials, cellulosic fibers can either be obtained from the respective plants, e.g. cotton, flax, ramie, jute, or by dissolution and regeneration of cellulosic material and left unmodified (regenerated cellulosics, such as viscose rayon, Tencel, etc.) or derivatized to result in modified regenerated cellulosic fibers such as cellulose acetate. The approximate chemical composition (without coloring matter, water solubles and moisture) of some important cellulosic fibers is listed in Table 2.1. For regenerated cellulosic fibers inexpensive sources for cellulose are identified. Examples include wood chips, cellulosic fibers too short for spinning, linters and others.

Table 2.1 Approximate chemical composition (%) of cellulosic fibers (Kraessig *et al.*, 1996; Lewin and Pearce, 1998)

Fiber	Cellulose	Hemicelluloses	Pectin	Lignin	Fat/wax
Seed hair fibers					
Cotton	92–95	5.7	1.2	0	0.6
Bast fibers					
Flax	62–71	16–18	1.8–2.0	2.0–2.5	1.5
Hemp	67–75	16–18	0.8	2.9–3.3	0.7
Ramie	68–76	13–14	1.9–2.1	0.6–0.7	0.3
Jute	59–71	12–13	0.2–4.4	11.8–12.9	0.5
Leaf fibers					
Sisal	66–73	12–13	0.8	9.9	0.3
Abaca	63–68	19–20	0.5	5.1–5.5	0.2
Nut husk fibers					
Coir	36–43	0.2	3–4	41–45	

Short-chain polymers of plant sugars are termed hemicellulose. They mainly consist of xylans, arabinogalactans and mannans (Gregory and Bolwell, 1999; Dey and Harborne, 1997). Pectinic substances are mainly calcium, magnesium and iron salts of polygalacturonic acid and the respective esters with a certain degree of branching (Dey and Harborne, 1997). Depending on the degree of esterification, they are insoluble in water, but can be removed by aqueous alkali or with the help of suitable enzyme systems. All natural cellulosic fibers, except for cotton, contain a certain portion of lignin, a complex polyphenolic network made of derivatives of phenylpropane. A detailed overview of the chemistry of lignin can be found in Hon and Shiraishi (2001).

Cellulose chains are bound by strong hydrogen bonds to form areas of high order (crystallinity) alternating with areas of less order. Strong sodium hydroxide and a few solvent systems are able to penetrate the crystalline areas; however water and most treatment solutions are only able to access the areas of low organization, resulting in swelling of the fiber. The moisture uptake of the fibers varies with the fiber type from approximately 6–7% (cotton, ramie) to 10–11% (jute, sisal; Lewin and Pearce, 1998) at standard conditions of 21°C and 65% relative humidity and can be modified to a certain extent by preparation treatments (mercerized cotton has a moisture regain of 8–12%).

Their ability to absorb water and swell has enabled dyeing and finishing of cellulosic fibers from aqueous solutions by exhaust procedures. Water-filled pores act as transportation ways for dyes, auxiliaries and other compounds to accessible functional groups in the interior of the cellulosic fiber.

2.21 Scanning electron microscopic picture of raw cotton fibers.

Increased temperature supports the process by accelerating the reaction rate; however, swelling already occurs at room temperature.

2.2.1.1 Natural cellulosics

Cotton

By far the most important textile cellulosic fiber is cotton. Cotton grows as unicellular fiber on seeds (seed hair fiber). A thin cuticle that mainly consists of waxes and pectins protects the outside of the fiber. The primary wall in mature fibers is only 0.5–1 μm thick and contains about 50% cellulose. Non-cellulosic impurities consist of pectins, fats and waxes, proteins and natural colorants. The secondary wall, containing approximately 92–95% cellulose, is built of concentric layers with alternating S- and Z-shaped twists. The layers consist of densely packed elementary fibrils, organized into microfibrils and macrofibrils. Bundles of fibrils, as well as individual arrangements of cellulose polymeric chains, are held together by strong hydrogen bonds. A hollow channel, the lumen, forms the center of the fiber. During plant growth it was filled with protoplasma. After maturing and harvest of the cotton fiber it dries and collapses giving the fiber a ribbon-like appearance with a kidney-shaped cross-section (Fig. 2.21).

2.2.1.2 Bast fibers

Flax

Besides cotton, fibers from flax (the term 'linen' generally describes fabric from flax fibers) are probably the second most important cellulosic fibers for apparel and household textiles. Suitable fibers are isolated from fibrous bast components by several steps. First, leaves and seeds are mechanically removed from the stalks (rippling), followed by the stripping of the bark (decortication). Subsequently attached woody compounds are chemically or enzymatically decomposed by retting. Scutching frees the coarse fiber bundles and during hackling the coarse bundles are separated into finer bundles. Flax fibers are usually not completely divided into single fibers but kept in small arrays of several individual fibers held together by gummy substances. In opposition to cotton, flax and all other bast fibers are multicellular, the cells being called ultimate cells or ultimates. Their cross-section is hexagonal with a lumen in the center. Along the fibers, characteristic cross markings, so-called nodes, are visible at the microscopic level. Purified flax fibers are approximately 70% crystalline. Owing to the presence of lignin and hemicelluloses, bast fibers have a higher moisture regain than cotton because these compounds add to the non-crystalline content of the fiber.

Ramie

Ramie is often blended with cotton for apparel because of its attractive luster. Fiber isolation from the plant follows similar procedures to those described for flax; however, ramie is embedded in a highly gummy pectinous bark and is hard to isolate by conventional retting processes. Enzymatic degumming using pectate lyase and xylanase has been explored (Bruhlmann *et al.*, 2000). Ramie is longer and coarser than flax. The cross-section of the fiber is multilobal with/without lumen. The crystallinity of ramie has been reported to be approximately 61% (Lewin and Pearce, 1998).

Hemp and jute

In many ways hemp fibers resemble flax and cotton fibers, but are less fine than flax. Their cross-section is uneven polygonal with rounded edges. Like flax and ramie, the lignin content is fairly low. The ratio of crystalline and less ordered regions in hemp is similar to that of flax.

The lignin content in jute is very high compared to the other bast fibers (see Table 2.1). The fiber has lower crystallinity, is hard to bleach to an acceptable whiteness and has a harsh hand. Retting of jute gives only a relatively small yield. Thus, the economic importance of jute for high quality textile materials has increasingly diminished over the years with synthetic fibers taking its place.

Leaf and nut husk fibers

Sisal, abaca (banana) and pina (pineapple) fibers are examples of leaf fibers; coir (coconut) belongs to nut husk or seed fibers. They are mostly used for cordage or industrial uses, in rare cases for decorative purposes (pina) and are not discussed here. Details on these fibers can be found in Lewin and Pearce (1998).

2.2.1.3 Common finishes for cellulosic fibers

Preparation finishing for natural cellulosic fibers includes desizing, scouring and bleaching (Trotman, 1984). Preparation can be performed in series or in a manner that combines two or all three steps. The purpose of scouring is to remove non-cellulosic impurities, oils and dirt, and the chemicals used for the process are hot aqueous alkali solution, often supported by detergents. Desizing can be achieved by hot water, acid or enzymatic hydrolysis, depending on the sizing compound used. For bleaching, oxidizing agents such as peroxides or chlorine-containing compounds are applied (see Section 2.1.2). Owing to the higher pectin content and the presence of lignin, bast fibers require stronger bleaching conditions. A scoured, desized and bleached fiber possesses excellent water absorbency and a high level of whiteness (exception: jute). Owing to the concerns outlined earlier, it is not surprising that milder processing alternatives are sought by the use of suitable enzyme systems.

Additionally, for high quality cotton goods mercerization is included in the preparation. Mercerization consists of a swelling process in 20–25% sodium hydroxide (Trotman, 1984). Sodium hydroxide in high concentration is capable of penetrating into the crystalline areas of the fiber and altering the cellulose I crystal lattice of native cotton to cellulose II, simultaneously affecting its pore structure and accessibility. If performed under tension of the textile material, the fibers obtain high luster owing to a rounder cross-sectional shape, increased tensile strength and dye uptake; if performed under slack conditions, enhanced water absorbency can be achieved.

Performance finishes and finishes adding functionality are optional. Generally, cellulosics show excellent comfort properties which are related to their moisture and heat transport capabilities. Their shortcomings are found in low wrinkle recovery, high shrinkage and high flammability. Finishes (Vigo, 1994) can be applied to address these properties to keep cellulosic fibers competitive with synthetics, such as polyester. The mechanism of crosslinking the structure of cellulosics and common chemicals used for easy-care properties are briefly outlined in Section 2.1.8; fluorochemicals to impart water- and stain-repellency are discussed in Section 2.1.6 and flame-retardant finishes are found in Section 2.1.9.

2.2.1.4 Regenerated cellulosic fibers

Viscose rayon

The process of regenerating cellulosic fibers in filament form from inexpensive cellulose sources has been modified many times over the past decades (Lewin and Pearce, 1998). In the viscose process, starting materials, such as wood chips, pulp or linters, are steeped in sodium hydroxide to form alkali cellulose, and allowed to age. During this process step considerable depolymerization takes place. The next processing stage involves the addition of carbon disulfide to form xanthate which is dissolved in dilute sodium hydroxide. After maturing, the viscose dope is filtered and extruded through spinnerets into an acidic regeneration bath where the fiber coagulates. Spinning speed, draw ratio, chemical auxiliaries and the composition of baths determine the properties of the resulting fiber (Kraessig *et al.*, 1996). Depending on the regeneration conditions the fibers show clearly defined skin–core structures. Skin and core exhibit different properties regarding, for example, crystallinity, accessibility and swelling.

The degree of crystallinity and orientation of regenerated cellulosic fibers strongly depend on the coagulation conditions and the applied draw ratio and are generally lower than those of natural cellulosic fibers. The basic structural unit of the regenerated fiber is the anhydroglucose unit (AGU), identical to that of natural cellulose. However, the DP amounts to approximately 400–600 only. The crystal lattice has been fully converted to cellulose II. Moisture uptake and water retention are typically higher than that of cotton. Dyes suitable for natural cellulosics can also be applied to regenerated cellulosics; however, care has to be taken because of the higher alkali sensitivity and the lower wet strength of some types of viscose rayon fibers.

Solvent-spun regenerated cellulosic fibers

The so-called solvent-spun fibers entered the market in the 1990s and the patent literature is vast, covering every step from dissolution and equipment to fiber modification (see Mulleder *et al.*, 1998, for example). The production of these fibers is based on the dissolution of cellulose in cyclic amines, such as *N*-methylmorpholine *N*-oxide/water (NMMO/H_2O), followed by a dry-jet–wet-spinning process (Marini and Brauneis, 1996).

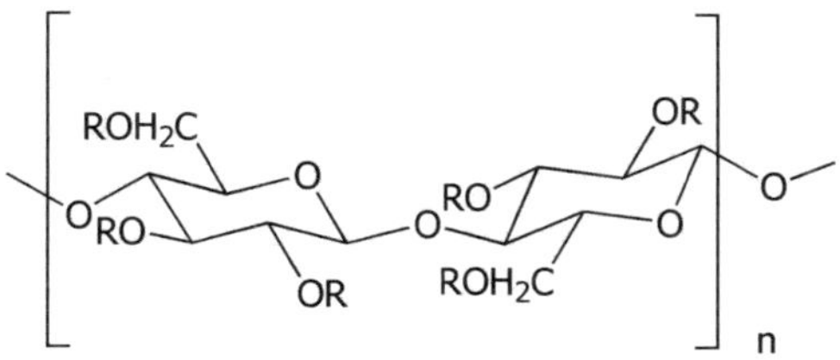

Cellulose triacetate (all R = $COCH_3$) and
cellulose acetate (2/3 R = $COCH_3$, 1/3 R = H)

2.22 Structure of cellulose acetate and triacetate.

Lyocell fibers, produced by Courtaulds under the trade name Tencel since 1993, have also been produced in European countries since mid-1997. Fibrillation of these fibers, although useful for fiber entanglement in nonwovens, has been a problem in other fabric forms, but is currently controlled by enzyme treatments or resin finishing (Bredereck *et al.*, 1997). NewCell (Krueger, 1994) and ALCERU fibers (Alternative Cellulosics Rudolphstadt), also obtained through dissolution of the raw materials in NMMO, are variations of Lyocell with slightly different fibrillation properties and as a result, slightly different finishing procedures. Because of the high recycle rate of the solvent NMMO and the fact that the substrate is complexed in the solution without chemical reaction, the production of these fibers is considered to be environmentally friendly.

A 'peach-skin effect' can be produced on solvent-spun fibers by mechanical fibrillation, followed by secondary fibrillation/defibrillation through enzymatic hydrolysis with fairly aggressive cellulases (Gandhi *et al.*, 2002). The result is a very fine fibrillic pile with soft, silk-like hand, and improved volume and appearance. This treatment is often accompanied by a silicone softener treatment (Breier, 1994).

2.2.1.5 Cellulose esters – acetate and triacetate

Three hydroxyl groups per AGU are available in cellulose that can be acetylated to 83–98.7% (Hatch, 1993) to form cellulose esters. Cellulose acetate consists of heterogeneous cellulose chains with 2.5 hydroxyl groups substituted by acetate groups on the average (Fig. 2.22). The DP of this fiber (250–300) is lower than that of viscose rayon, and the degree of orientation and crystallization is very low with fewer intermolecular hydrogen bonds. While acetate fibers are still fairly hydrophilic with a moisture regain quite close to that of cotton, triacetate exhibits the properties of a hydrophobic fiber. To increase its low crystallinity after extrusion, triacetate has to be heat-set, a process routinely applied to most synthetic fibers. During the heat-setting process, the polymeric chains are arranged in closely packed arrays to form a highly crystalline structure. Acetate, on the other hand, cannot be heat-set. As with non-modified regenerated cellulosics, various cellulose acetate and triacetate types and manufacturing methods exist, resulting in fibers with a wide range of properties (Lewin and Pearce, 1998).

2.2.1.6 Possibilities for enzyme applications for cellulosic substrates

Published research on enzymatic hydrolysis of cellulosic materials is very extensive. Suitable enzymes for cellulosic substrate surface finishing are cellulases which are commercially available with various pH and temperature activity profiles. Their properties and modes of action are covered elsewhere in this book. Mild treatments lead to surface polishing by removing small fiber fibrils on the surface, rendering the textile material softer and improving color brilliance. Harsher conditions can lead to higher weight and tensile strength losses and eventually to the complete fiber breakdown. Enzymatic finishing processes may be performed before or after coloration as well as before or after selected chemical finishing procedures.

Commonplace in denim finishing today is the biostoning process to give jeans a worn and washed appearance, replacing the pumice stones that were traditionally used. Various approaches have been taken, including cellulases with or without a reduced quantity of pumice stones (Klahorst *et al.*, 1994), mixtures of amylases and cellulases (Uhlig, 1998) and laccases (Mueller and Shi, 2001).

Enzymatic scouring has generated a great deal of interest in the light of cost savings and growing environmental concerns. Pectinases, cellulases, proteases and lipases have all been investigated with respect to their effectiveness in removing non-cellulosic impurities and increasing the wettability of the textile material (Roessner, 1995; Buschle-Diller *et al.*, 1998; Takagishi *et al.*, 2001; Traore and Buschle-Diller, 2000; Waddell, 2002). More recently, efforts to include enzymatic bleaching with glucose oxidases and peroxidases have also been reported (Buschle-Diller *et al.*, 2001; Tzanov *et al.*, 2002) with the glucose oxidase in free form or immobilized on a support material.

2.2.2 Protein fibers

Protein fibers for textile uses can be divided into animal hair fibers, such as wool (defined as fibers from various breeds of domesticated sheep) and speciality hair (all other animal hair fibers, such as mohair, cashmere, alpaca, angora, etc.), and animal secretion fibers (silk).

2.2.2.1 *Wool and animal hair fibers*

Fibers obtained from sheep and other animals vary significantly in length and fiber fineness, shape, pigmentation, crimp and surface properties. Within one breed, variations occur that depend on the age and the nutritional condition of the animal and on the part of the body from which the fibers were acquired.

All animal hair fibers have some common features. The exterior of wool and other animal fibers consists of flat overlapping cuticle cells ('scales') that protect long spindle-shaped cortex cells in the interior. The cortical cells surround the innermost cells with vacuoles, the medulla, which might be missing in some types of finer animal hair fibers. Each cuticle cell is made up of three layers, epicuticle, exocuticle and endocuticle. Cortex cells consist of ortho, para and meso cells. Each layer of the cuticle and the cortex cells differ in cystine and isodipeptide content. The cell membrane complex, which is made of proteins with low crosslink density and lipids, fuses the different layers of the fiber together.

Keratin, the wool protein, is built up from a total of 24 major amino acids with different functionalities (Fig. 2.23). Amino acids with acidic character include aspartic acid, glutamic acid, asparagine and glutamine; basic character is introduced by arginine and lysine, histidine and tryptophan. The wool fiber is thus amphoteric and internal salt bridges can be formed. At the isoelectric point (pH 4.8–4.9), the number of anionic groups equals the number of cationic groups and the fiber is in its most stable form. Besides ionic interactions, amino acids containing hydroxyl groups (serine, threonine and tyrosine) are capable of forming intramolecular hydrogen bonds to add stability to the wool fiber. Additionally, unlike other proteins, several amino acids carry sulfur-containing side chains – cysteine, cystine, cysteic acid, methionine, thiocysteine and lanthionine – with cystine accounting for the highest weight percentage of sulfur in the form of disulfide bridges. These disulfide bridges which are formed between two polypeptide chains significantly influence fiber stability. The remaining amino acids do not have any specific reactivity. They include glycine, leucine, proline, valine, alanine, isoleucine and phenylalanine. In summary, stability of the wool fiber is thus achieved not only through the formation of salt bridges and hydrogen bonds, but also via disulfide bridges between two cystine residues, covalent bonding between glutamic acid and lysine residues, and hydrophobic interactions between non-polar amino acid side groups (Zahn and Hoffmann, 1996).

Each polypeptide is twisted in the form of an α-helix (right-handed screw). Two such helices are joined together to form a left-handed coiled assembly. With the hydrophobic side chains pointing to the outside, the chains are held together tightly through intramolecular hydrogen bonds. By

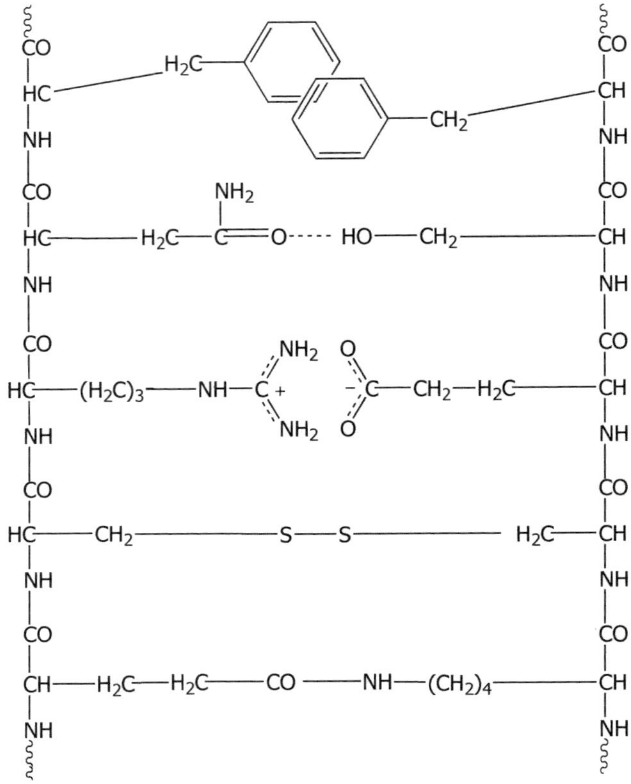

2.23 Wool keratin.

coming close enough, crystalline areas that alternate with amorphous regions are formed.

Fatty acids (2%) and proteinaceous compounds are found in the cuticle surface and the cell membrane complex that holds cuticle and cortex cells together. These fatty acids include palmitic, stearic and oleic acid as free fatty acids, 18-methyleicosanoic acid bound to proteins, cholesterol and cholesterol sulfate, as well as polar lipids, such as ceramides and cerebrosides.

The fatty acids provide a hydrophobic barrier at the fiber surface while the interior of the fiber is hydrophilic. The moisture regain at 65% relative humidity and 21°C is approximately 12–15%. The uptake of liquid water is accompanied by considerable radial swelling, especially above and below the isoionic point. With excessive positive or negative charge the polypeptide chains electrostatically repel each other, thus allowing for increased swelling accompanied by decreased mechanical stability. In case friction or agitation is present during the wetting process, the scales interlock and form a felted structure (felting shrinkage, see below).

Owing to their complex composition, protein fibers can undergo reactions occurring at covalent bonds including the polypeptide backbone, and within the side chains. Chemical degradation occurs with moist heat, under alkaline conditions and with strong mineral acids by hydrolysis of polypeptide bonds as well as degradation of some of the side chains, liberating hydrogen sulfide, ammonia and other decomposition products. Oxidizing and reducing agents predominately attack the disulfide bridges. Splitting and reformation of disulfide bridges under controlled conditions, however, allows for setting of wool for stabilization reasons. Amino groups of the side chains are major dye sites for acid dyes, while anionic groups can form covalent bonds with specifically developed reactive dyes (for more details on wool dyeing see, for example, Bearpak *et al.*, 1986 and Lewis, 1992).

2.2.2.2 *Common chemical finishes for protein fibers*

Raw animal fibers contain high amounts of grease, suint and vegetable matter with the average amounts varying with animal rearing conditions. Although wool grease is easily solubilized in organic solvents, hot water or aqueous alcohol is usually the scouring method of choice to remove suint and soil impurities simultaneously (Zahn and Hoffmann, 1996). Great care has to be taken not to cause felting shrinkage through mechanical movement of the fibers in the scouring bath. The scoured fibers are treated in a way so as to retain a minor amount of grease (about 0.5%). If unacceptable quantities of cellulosic matter contaminate the fibers, an addition carbonization treatment can be performed which constitutes short exposure to sulfuric acid with heating or treatment with cellulases.

As mentioned, a disadvantage of animal fibers is their tendency to cause felting shrinkage, caused by the scale structure of the cuticle. Anti-felting finishes are targeted to reduce the rough surface, either by partial removal of the scales by chemical treatment, by coating the scales with a polymer, or by preventing their contact through spot welding by deposition of polymer aggregates that keep the fibers at a fixed distance from each other (Vigo, 1994). Chemical treatments involve reducing agents, solvents or oxidizing reagents, such as chlorine, peroxysulfuric acid or permanganate. Because the conditions of this treatment are fairly harsh and might lead to fiber damage, a later development combines a milder chlorination treatment with hypochlorite and a polyamide coating (Hercosett®). Such anti-felting finishes produce washable wool, although they affect the dyeing behavior, hand and other properties. Newer developments focus on the application of plasma (Hoecker, 1997) and enzymatic processes for wool fibers (El-Sayed *et al.*, 2002).

Enzymes have been used to support the descaling process with the goal of improving shrink resistance and hand of wool (Heine and Hoecker, 1995; Heine *et al.*, 1998; Galante *et al.*, 1998; Breier, 2000) and for bleaching and scouring purposes (Levene, 1997; Brahimi-Horn *et al.*, 1990). Most research work on softening and diminishing the cuticle scale structure has been concerned with finding suitable proteases that would not significantly weaken the fiber otherwise. Proteases of plant origin, such as papain, as well as from other sources, have been explored, either in combination with a chemical treatment, e.g. chlorination, or by themselves. For scouring, various lipases and esterases have been studied.

2.2.2.3 Silk fibers

Silk as a useful animal secretion fiber is obtained from the domesticated moth (*Bombyx mori*) feeding solely on mulberry leaves, and from the wild tussah varieties (*Antheraea pernyi* and *A. mylitta*), feeding on oak or castor leaves. Spider silks of the Arachnida family have been explored, so far mostly for research purposes. Silk is extruded from the silk-producing glands of the larvae as a double filament, made of fibroin, held together by a cementing layer of sericin and solidified in air. The cross-section of cultivated silk is irregular trilobal with rounded edges, while wild silk is flatter, wedge-shaped with less sericin.

Silk consists of 18 amino acids with glycine, alanine, serine, and to a minor extent, tyrosine, making up more than 90 mol% of the fibroin (Zahn, 1993). In cultivated silk, glycine amounts to almost 45%, while in wild silk alanine predominates. The amount of cystine is very small in both types. Overall, the total number of acidic groups is two to three times that of basic groups (isoelectric range pH 4–5). The fibers are highly oriented. Owing to the small quantity of bulky side groups, the polypetide chains permit the formation of a β-pleated sheet structure instead of a helix as is the case for wool (Fig. 2.24). The sheets can stack and form crystalline regions with intramolecular hydrogen bonding between the sheets.

Sericin, which makes up 17–25% of the fiber weight, significantly differs from fibroin. Its major amino acids are glycine, serine and aspartic acid in cultivated silk, and glycine, serine, threonine and aspartic acid in wild silk with serine occurring in the greatest amount in both silk types. The ratio of non-polar to polar amino acid residues is 1:3, with approximately 60% hydroxyl groups, 30% acidic and 10% basic groups (isoelectric point at pH 4.0). The cystine content is slightly higher than in fibroin. Sericin also contains about 1.5% fats and waxes and about 1% mineral compounds.

The virtual absence of disulfide bridges renders the silk fiber more sensitive to acids but less sensitive to alkalis than wool fibers. Large amounts of polar amino acids account for the hydrophilicity of the fiber. Under standard conditions of 65% relative humidity and 21°C, the moisture regain of silk is approximately 10–11%. Silk can absorb considerable quantities of salt solutions, a property that is used in the process of weighting of silk to

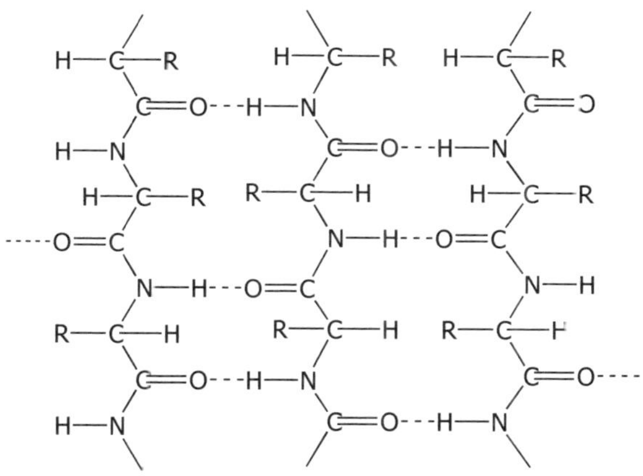

2.24 Chemical structure of silk (for R see text).

correct for the weight lost during degumming (see below). Acid dyes are the dyes of choice for silk fibers today, because the basic dyes as used in earlier years proved to have rather limited light fastness. Tussah silk is more resistant against chemicals and can also be dyed with reactive dyes under alkaline conditions. Washable silk usually refers to silk coated with a protective polymer (Tsukada *et al.*, 2001).

The separation of the silk double filaments (degumming) by removal of sericin has conventionally been achieved by immersion of the fibers in alkaline solution and soap. This process results in considerable weight loss (up to 20%). Enzymatic degumming with bacterial or fungal proteases leads to improved dyeing and hand properties with slightly lower weight loss (Uhlig, 1998; Gulrajani *et al.*, 1996).

2.2.3 Synthetic fibers

Generally, natural fibers are true substrates in enzymatic processes. Synthetic fibers have been explored in the context with their support properties for enzyme immobilization and for special applications such as biosensors or membranes. A few selected synthetic fibers have also been subjected to enzymatic modifications in the form of textile substrates, the most frequently studied probably being polyester (Yoon *et al.*, 2002). Therefore only major synthetic fibers will be briefly discussed below. If more detailed information is needed, the reader is referred to a general fiber science book, such as Lewin and Pearce (1998).

All synthetic fibers are petroleum derivatives. They are designed in the chemical laboratory and created to achieve the most favorable properties at reasonable cost. Production conditions determine the composition, DP,

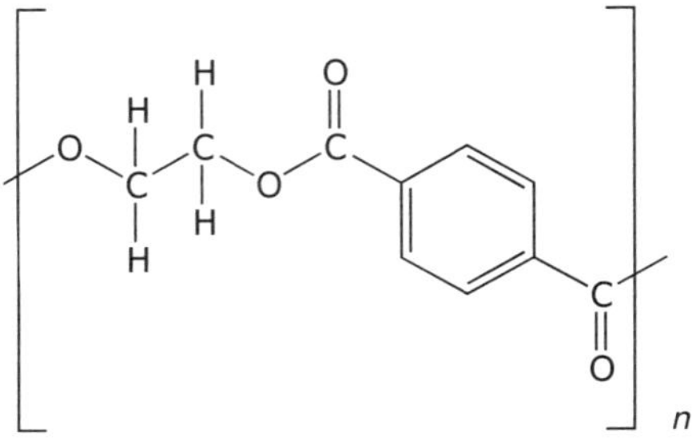

2.25 Structure of PET.

fiber diameter and shape, mechanical properties, etc. of the final fiber. Synthetic fibers are generally formed as continuous strands of filaments, which may be cut into staple lengths if necessary. With advances in polymer synthesis, engineering and fiber formation methods, more and more custom-made fibers are entering the market place.

2.2.3.1 Polyester (poly(ethylene terephthalate))

By a generic definition, fibers composed of at least 85% by weight of an ester of a substituted aromatic (or aliphatic) carboxylic acid are termed polyesters (Hatch, 1993). Polyesters with a sufficient DP are generally made by reaction of diols with dicarboxylic acids. The most important representative of this category is poly(ethylene terephthalate) or PET (Fig. 2.25).

The draw ratio and processing history determine the degree of orientation and also influence the crystallinity of the fiber. Usually, highly crystalline areas alternate with regions of low crystallinity. Tensile strength, extension at break and initial modulus are directly related to the ratio of ordered to less ordered regions and the degree of orientation. In comparison with other synthetic fibers, PET belongs to the stronger (in the range of nylon and polypropylene) and stiffer fibers, although the different PET types vary according to their manufacturing conditions. Owing to the high tensile strength of PET, pilling – the formation of small fiber pills on the surface of a polyester fabric – presents a problem, especially when PET is blended with other, less strong fibers.

Polyester can be hydrolyzed under alkaline conditions. The rate of hydrolysis is very low without a catalyst and occurs only at the surface. This process that can be used to etch the surface increases the hydrophilicity of the fiber and alters its hand properties. Inorganic acids with catalysts that have sufficient diffusive capabilities, organic acids (e.g. dichloroacetic acid), amino compounds, such as ammonia, primary and secondary amines, also hydrolyze polyester.

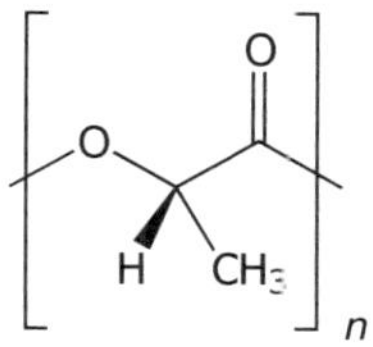

2.26 L-Poly(lactic acid), PLA.

PET is a hydrophobic fiber with maximum moisture regain of only 1% at 100% relative humidity. Until the development of disperse dyes, dyeing of polyester was difficult. Disperse dyes with very low water solubility can sublime into the fiber by heat (thermosol process, thermofixation), be applied with heat/pressure or with the help of carriers by an exhaust process (see Section 2.1.1). Basic-dyeable modified polyesters, copolymerized with units containing sulfonate groups, are also commercially available.

2.2.3.2 *Biopolyesters*

A newly emerging field comprises biopolyesters, encompassing polyesters produced by biocatalytic means, such as enzyme-catalyzed polymerizations, as well as polymers from biological origins or from renewable resources (Scholz and Gross, 2000). Poly(hydroxyalkanoates), PHAs, are the only biopolymers that are completely synthetized by microorganisms. They present energy and nutrient storage for the microorganisms in case the environment changes triggering the enzymatic breakdown of the biopolyesters.

The most important examples of compounds generated by biocatalytic routes are poly(lactic acid), PLA, and poly(glycolic acid), PGA, as well as their copolymers (Fig. 2.26). These polyesters exhibit high tensile strength and are non-toxic. Crystallinity, orientation and moisture sorption capabilities of the fibers are controlled by the production conditions. Such compounds are fully biodegradable and can be subjected to enzymatic modification.

2.2.3.3 *Polyamides*

In polyamides, the structural units are connected by amide groups. In generic terms, aliphatic polyamides are called nylons, aromatic polyamides are called aramids. Aramids will not be discussed here. Polyamides are formed either by condensation of diamines and diacids or by ring-opening polymerization of lactams. In aliphatic polyamides, nylons are named as (A,B), where the numbers signify the number of carbon atoms of the diamine and the diacid linked together (for example, nylon 6,10). If a single

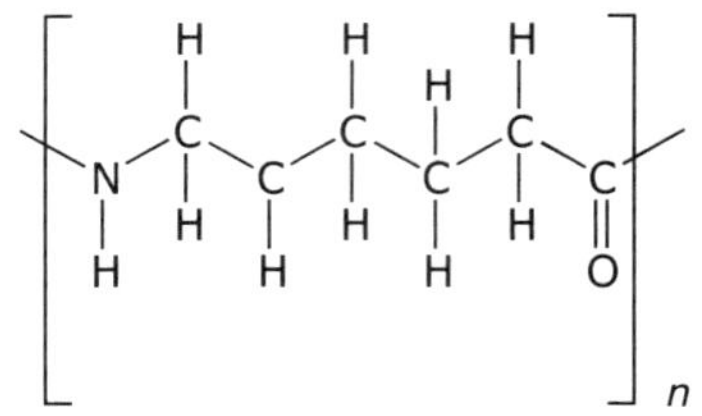

(a)

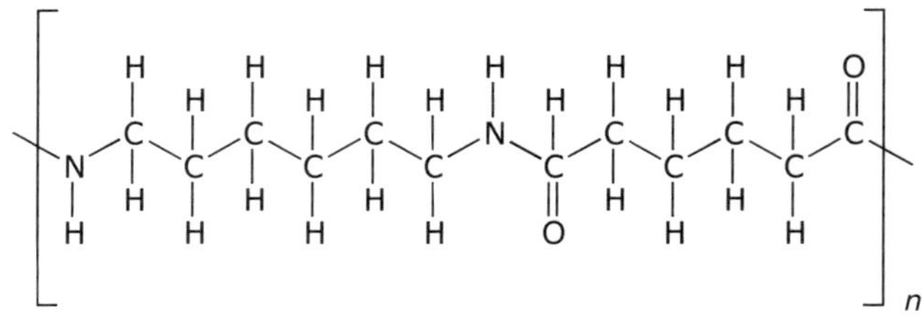

(b)

2.27 (a) Nylon 6 and (b) nylon 6,6.

number is given, it refers to the number of carbon atoms of a nylon produced by ring-opening polymerization (e.g. nylon 6 from ε-caprolactam). A large array of products is available, including various copolymers; however, the aliphatic polyamides of most economic significance are nylon 6, nylon 6,6 and nylon 6,10 (Fig. 2.27).

In general, both polymerization methods lead to a mixture of polymers with various molecular weights. Apart from the DP and the molecular weight distribution, the nature and number of end groups is important for chemical reactivity and dyeing purposes. Nylons can be produced with approximately equal numbers of acidic and amino end groups (regular type), mostly acidic (acid-dye resistant nylon) or mostly amino end groups (deep dyeing nylons). Zigzagging carbon segments line up closely between the end groups. They are held together by van der Waals forces and intramolecular hydrogen bonding through amide links to form sheet-like arrangements. By tight stacking of these sheets, crystalline regions are created alternating with less organized areas that do not have distinct boundaries. Drawing may cause the development of crystalline areas in those less ordered regions. The particular molecular arrangement in nylons results in high tensile strength, elongation and elastic recovery upon application of stress.

Nylons are dyeable with disperse dyes (shade range is limited) or with acid dyes under mild acidic conditions. Aqueous acids (below pH 3) as well as bases cause the rupture of the polymer backbone. In the case of acid dyeing, dye molecules only attach to available amino end groups, thus shade depth is determined by the ratio of negatively charged groups of the dye molecule to positively charged end groups in the fiber. Besides hydrolysis

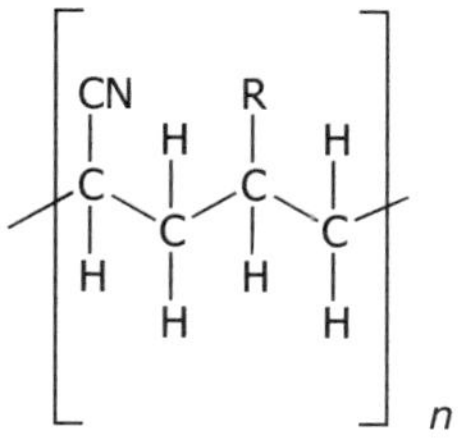

2.28 Polyacrylonitrile (homopolymer, R=CN and copolymer, R=COOH, SO_3H, etc.).

by chemical means, nylons are fairly susceptible to heat and light, manifested by decreased tensile strength and a yellowed appearance of the fiber. Commercially available fibers may thus contain antioxidants to reduce light sensitivity or copper(I) salts to improve thermal resistance.

2.2.3.4 Polyacrylonitriles

Commercial acrylic fibers contain at least 85% acrylonitrile copolymer (modacrylic fibers between 35 and 85%) combined with one or more other monomers (Fig. 2.28). The homopolymer (100% acrylonitrile) is difficult to process and dye, and thus is only made for industrial applications. The comonomers in acrylic fibers are selected with the purpose of giving the fiber specific properties, such as dyeability (sodium methallyl sulfonate, sodium sulfophenyl methallyl ether, etc.), modified fiber morphology (vinyl acetate, methyl acrylate, etc.) or flame retardancy (vinyl bromide, vinyl chloride, etc.). Copolymerization with different comonomers has opened up a venue to innumerable speciality products. Bicomponent fibers from acrylics of different composition with distinct properties add to these possibilities. For example, side-by-side bicomponent fibers exhibiting different shrinkage ratios impart crimp upon exposure to heat or wet conditions.

Commercial acrylic fibers are produced by free radical processes which make stereoregularity and thus crystallinity problematic. Nevertheless, with the help of spectroscopic methods it can be shown that acrylic fibers contain crystalline and less ordered areas with strong connections between the two phases. Intermolecular dipolar bonding accounts for relative stiffness in the lower ordered regions. The introduction of a comonomer generally reduces the crystallinity and the melting point. Thus, tensile strength and breaking elongation are lower compared with other synthetic fibers, such as polyester or nylon. Moisture regain is fairly low, depending to a certain extent on the nature of the comonomer.

Acrylic fibers with introduced negative groups can be dyed with basic (cationic) dyes under carefully controlled conditions. Dyeing is usually performed in the presence of a retarder to decrease the rate of the dyeing process for uniform shade reproduction. Finishing processes for polyacrylonitrile are limited since desirable properties can be more easily incorporated by copolymerization or by modification on the fiber level. For example, highly absorbent fibers are made by inclusion of a hydrophilic comonomer which is subsequently removed by hydrolysis.

2.3 References

Abadulla E., Tzanov T., Costa S., Robra K.-H., Cavaco-Paulo A. and Guebitz G.M. (2000) 'Decolorization and detoxification of textile dyes with a laccase from *trametes hirsute*', *Appl. Environ. Microbiol.*, **66** (8), 3357–3362.

Bearpak I., Marriott F.W. and Park J. (1986) *A Practical Introduction to the Dyeing and Finishing of Wool Fabrics*, Society of Dyers and Colourists, Bradford, UK.

Brahimi-Horn M.C., Guglielmino M.L., Gaal A.M. and Sparrow, L.G. (1990) 'Potential uses of enzymes in early processing of wool', *Proc. 8th Int. Wool Textile Res. Conf.*, **3**, 205–214.

Bredereck K., Schulz F. and Otterbach A. (1997) 'Fibrillation propensity of lyocell and the influence of reactive dyeings', *Melliand Internat.*, **78**, 217.

Breier R. (1994) 'Die Veredlung von Lyocellfasern–Ein Erfahrungsbericht', *Lenzinger Ber.*, **4**, 99–1001.

Breier R. (2000) 'Lanazym process: purely enzymatic antifelt finishing of wool', *Melliand Textilber.*, **81** (4), E77–E79; **298**, 300–302.

Broze G. (ed.) (1999) *Surfactant Science Series*: *Handbook of Detergents, Part A, Properties*, Marcel Dekker, New York, Vol. 82.

Bruhlmann F., Leupin M., Erismann K.H. and Fiechter A. (2000) 'Enzymatic degumming of ramie bast fibers', *J. Biotechnol.*, **76** (10), 43–50.

Buschle-Diller G., El Mogahzy Y., Inglesby M.K. and Zeronian S.H. (1998) 'Effects of scouring with enzymes, organic solvents, and caustic soda on the properties of hydrogen peroxide bleached cotton yarn', *Textile Res. J.*, **68** (12), 920–929.

Buschle-Diller G., Yang X.D. and Yamamoto R. (2001) 'Enzymatic bleaching with glucose oxidase', *Textile Res. J.*, **71** (5), 388–394.

Call H.P. and Muecke I. (1997) 'History, overview and application of mediated lignolytic systems, especially laccase-mediator-systems (Lignozym-Process)', *J. Biotechnol.*, **53**, 163–202.

Campos R., Cavaco-Paulo A., Robra K.-H., Schneider M. and Guebitz G. (2001) 'Indigo degradation with laccases from *polypous sp.* and *sclerotium rolfsii*', *Textile Res. J.*, **71** (5), 420–424.

Colour Index (1999) 3rd Edition on CD, Society of Dyers and Colourists and American Association Textile Chemists Colorists, Bradford, UK.

Datyner A. (1993) 'Interactions between auxiliaries and dyes in the dyebath', *Rev. Prog. Color.*, **23**, 40–50.

De Baets S., Vandamme E.J. and Steinbuechel A. (eds) (2002) *Biopolymers, Vol 6: Polysaccharides II, polysaccharides from eukaryotes*, Wiley-VCH, Weinheim, FRG.

Dey P.M. and Harborne J.B. (eds) (1997) *Plant Biochemistry*, Academic Press, London.

El-Sayed H., Hamed R.R., Kantouch A., Heine E. and Hoecker H. (2002) 'Enzyme-based feltproofing of wool', *AATCC Rev.*, **2** (1), 25–28.
Flick E.W. (1993) *Industrial Surfactants*, 2nd Edition, Noyes Publications, Park Ridge NJ.
Fukamizo T. (2000) 'Chitinolytic enzymes: catalysis, substrate binding, and their application', *Curr. Protein Peptide Sci.*, **1** (1), 105–124.
Galante Y.M., Foglietti D., Innocenti R., Ferrero F. and Monteverdi R. (1998) 'Interaction of subtilisin-type protease with merino wool fibers', in *Enzyme Applications in Fiber Processing*, eds Eriksson K.E.L. and Cavaco-Paulo A., Chapter 24, ACS Symposium Series 687, American Chemical Society.
Gandhi K., Burkinshaw S.M., Taylor J.M. and Collins G.W. (2002) 'A novel route for obtaining a "peach-skin effect" on lyocell and its blends', *AATCC Rev.*, **2** (4), 48–52.
Glasser W.G., McCartney B.K. and Samaranayake, G. (1994) 'Cellulose derivatives with a low degree of substitution. Part 3. The biodegradability of cellulose esters using a simple enzyme assay', *Biotechnol. Prog.*, **10** (2), 214–219.
Gregory A. and Bolwell G.P. (1999) 'Hemicelluloses', in *Comprehensive Natural Products Chemistry*, ed. Pinto B.M., Vol. 3, Chapter 3, Elsevier Science BV, Amsterdam.
Gulrajani M.L., Gupta S.V., Gupta A. and Suri M. (1996) 'Degumming of silk with different protease enzymes', *Indian J. Fibre Textile Res.*, **21** (4), 270–275.
Guthrie J.T. (1990) 'Polymeric colorants', *Rev. Prog. Color Related Topics*, **20**, 40–52.
Hatch K.L. (1993) *Textile Science*, West Publishing, Minneapolis/Saint Paul.
Heine E. and Hoecker H. (1995) 'Enzyme treatments for wool and cotton', *Rev. Prog. Color Related Topics*, **25**, 57–63.
Heine E., Hollfelder B., Lorenz W., Thomas H., Wortmann G. and Hoecker H. (1998) 'Enzymes for wool modification', in *Enzyme Applications in Fiber Processing*, eds Eriksson K.E.L. and Cavaco-Paulo A., Chapter 23, ACS Symposium Series 687, American Chemical Society.
Heinze T. (1998) 'New ionic polymers by cellulose functionalization', *Macromol. Chem. Phys.*, **199**, 2341–2364.
Helle S.S., Duff J.B. and Cooper D.G. (1993) 'Effect of surfactants on cellulose hydrolysis', *Biotech. Bioeng.*, **42**, 611–617.
Hoecker H. (1997) 'Wool, current challenges, attempts and solutions', *Textilveredlung*, **32** (7/8), 154–155.
Hon D.N.-S. and Shiraishi N. (eds) (2001) *Wood and Cellulosic Chemistry*, 2nd Edition, Marcel Dekker, New York.
Kaya F., Heitmann J.A. and Joyce T.W. (1995) 'Influence of surfactants on the enzymatic hydrolysis of xylan and cellulose', *Tappi*, **78** (10), 150–157.
Klahorst S., Kumar A. and Mullins M.M. (1994) 'Optimizing the use of cellulase enzymes', *Textile Chem. Color.*, **26** (2), 13–18.
Kraessig H., Steadman R.G., Schliefer K. and Albrecht W. (1996) 'Cellulose', in *Ullmann's Encyclopedia of Industrial Chemistry*, Vol. A. 5, VCH Verlagsgesellschaft, Weinheim, Germany.
Krueger R. (1994) 'Cellulosic filament yarn from the NMMO process', *Chemiefasern/Textilind.*, **44** (1–2), 24–7.
Kumar M.N.V.R. (2000) 'A review of chitin and chitosan applications', *React. Funct. Polym.*, **46**, 1–27.

Levene R. (1997) 'Enzyme-enhanced bleaching of wool', *J. Soc. Dyers Color.*, **113**, 206–209.
Lewin M. and Pearce E.M. (eds) (1998) *Handbook of Fiber Chemistry*, 2nd Edition, Marcel Dekker, New York.
Lewin M. and Sello S.B. (eds) (1983a) *Handbook of Fiber Science and Technology, Vol. 1, chemical processing of fibers and fabrics: fundamentals and preparation, Part A and B*, Marcel Dekker, New York.
Lewin M. and Sello S.B. (eds) (1983b) *Handbook of Fiber Science and Technology, Vol. II, chemical processing of fibers and fabrics: functional finishes, Part A and B*, Marcel Dekker, New York.
Lewis D.M. (ed.) (1992) *Wool Dyeing*, Society of Dyers and Colourists, Bradford, UK.
Lin M. and Zheng L. (2002) 'Boron compounds as flame retardants and their synergy with phosphorous', *AATCC Rev.*, **2** (2), 30–33.
Marini I. and Brauneis F. (1996) 'Lenzing-Lyocell. A cellulosic fiber with new properties', *Textilveredlung*, **31** (9/10), 182–187.
Miles L.W.C. (ed.) (1994) *Textile Printing*, 2nd Edition, Society Dyers and Colourists, Bradford UK.
Mueller M. and Shi C. (2001) 'Laccase for denim processing', *AATCC Rev.*, **1** (7), 4–5.
Mulleder E., Schrempf Ch., Ruf H. and Feilmair W. (1998) *Solvent-spun regenerated cellulosic microfibers with fine denier*, PCT Int Appl, CAN **130**, 82801.
Philipp B. and Stscherbina D. (1992) 'Enzymatic degradation of cellulosic derivatives in comparison to cellulose and lignocellulose', *Papier*, **46** (12), 710–722.
Raheel M. (1998) 'Single-step dyeing and formaldehyde-free durable press finishing of cotton fabric', *Textile Res. J.*, **68** (8), 571–577.
Reife A. and Freeman H.S. (1996) *Environmental Chemistry of Dyes and Pigments*, John Wiley, New York, pp 205–207.
Rivlin J. (1992) *The Dyeing of Textile Fibers – Theory and Practice*, Philadelphia College of Textiles and Science, Philadelphia, PE.
Roessner U. (1995) 'Enzyme in der Baumwollvorbehandlung', *Textilveredlung*, **30**, 82–89.
Scholz C. and Gross R.A. (eds) (2000) *Polymers from Renewable Resources, Biopolyesters and Biocatalysis*, ACS Symposium Series 764, Oxford University Press.
Schweppe H. (1993) *Handbuch der Naturfarbstoffe*, Ecomed Verlagsgesellschaft, Landsberg, Austria.
Shore J. (ed.) (1990a) *Colorants and Auxiliaries, Organic Chemistry and Application Properties, Volume 1, Colorants*, Society Dyers and Colourists, Bradford, UK.
Shore J. (ed.) (1990b) *Colorants and Auxiliaries, Organic Chemistry and Application Properties, Volume 2, Auxiliaries*, Society Dyers and Colourists, Bradford, UK.
Shore J. (1995) *Cellulosic Dyeing*, Society Dyers and Colourists, Bradford, UK.
Shugar G.J. and Ballinger J.T. (1990) *Chemical Technicians' Ready Reference Handbook*, 3rd Edition, McGraw-Hill, New York.
Slate P.E. (1998) *Handbook of Fiber Finish Technology*, Marcel Dekker, New York.
Takagishi T., Yamamoto R., Kikuyama K. and Arakawa H. (2001) 'Design and application of continuous bio-scouring machine', *AATCC Rev.*, **1** (8), 32–34.
Traore M.K. and Buschle-Diller G. (2000) 'Environmentally friendly scouring processes', *Textile Chem. Color. Am. Dyestuff Rep.*, **32** (12), 40–43.

Trotman E.R. (1984) *Dyeing and Chemical Technology of Textile Fibres*, Wiley, New York.

Tsukada M., Arai T., Winkler S., Freddi G. and Ishikawa H. (2001) 'Physical properties of silk fibers grafted with vinyltrimethoxysilane', *J. Appl. Polym. Sci.*, **79** (10), 1764–1770.

Tzanov T., Costa S., Calafell M., Guebitz G. and Cavaco-Paulo A. (2000) 'Enzymes for cotton fabrics preparation and recycling of waste waters for dyeing', *Colourage Ann.*, 65–68, 70–72.

Tzanov T., Costa S., Calafell M., Guebitz G. and Cavaco-Paulo A. (2002) 'Hydrogen peroxide generation with immobilized glucose oxidase for textile bleaching', *J. Biotechnol.*, **93**, 87–94.

Ueda M., Koo H. and Wakida T. (1994) 'Cellulase treatment of cotton fabrics, Part II: Inhibitory effect of surfactants on cellulase catalytic reaction', *Textile Res. J.*, **64** (10), 615–618.

Uhlig H. (1998) *Industrial Enzymes and their Applications*, John Wiley, New York.

Vigo T. (1994) *Textile Processing and Properties*, Elsevier, Amsterdam.

Waddell R.B. (2002) 'Bioscouring of cotton: Commercial applications of alkaline stable pectinases', *AATCC Rev.*, **2** (4), 28–30.

Yoon M.Y., Kellis J. and Poulose A.J. (2002) 'Enzymatic modification of polyester', *AATCC Rev.*, **2** (6), 33–36.

Zahn H. (1993) 'Silk', in *Ullmann's Encyclopedia of Industrial Chemistry*, Vol. A. 24, VCH Verlagsgesellschaft, Weinheim, Germany.

Zahn H. and Hoffmann R. (1996) 'Wool', in *Ullmann's Encyclopedia of Industrial Chemistry*, Vol. A. 28, VCH Verlagsgesellschaft, Weinheim, Germany.

Zollinger H. (1991) *Color Chemistry*, 2nd Edition, Verlag Chemie, Heidelberg, FRG.

Zeronian S.H. and Inglesby M.K. (1995) 'Bleaching of cellulose with hydrogen peroxide', *Cellulose*, **2**, 265–272.

3
Catalysis and processing

ARTUR CAVACO-PAULO
University of Minho, Portugal
GEORG GÜBITZ
Graz University of Technology, Austria

The function and application of enzymes used in textile processing are discussed in this chapter which is composed of four parts: basic thermodynamics and enzyme kinetics, function of textile processing enzymes, homogenous and heterogeneous catalysis and important applications of enzymes in textile wet processing.

The first part on thermodynamics and kinetics of enzymes describes basic thermodynamics of chemical reactions, including the concepts of free energy, collision theory and catalysed reactions. After a general introduction to enzymes, the second part gives an overview of the catalytic mechanisms of enzymes used in textile processing including amylases, cellulases, pectinolytic enzymes, esterases, proteases, nitrile hydrolysing enzymes, catalases, peroxidases and laccases.

Substrate–enzyme interactions at the active sites of these enzymes belonging to different classes are discussed and parameters influencing the reactions are listed.

Subsequently, the function of enzymes in homogenous and heterogenous enzymatic reactions is discussed more in detail and important models such as those of Michealis–Menten and Briggs–Haldane are presented. Furthermore, parameters influencing the performance of enzymes such as enzyme stability and the presence of inhibitors are discussed based on models. The relevance of these models for the development of industrial processes is shown.

In the last part of this chapter, an introduction is given to classical textile wet processing followed by a description of successful enzyme applications in textile processing such as enzymatic desizing, degradation of hydrogen peroxide in bleaching effluents by catalases, cellulase finishing and enzymes in detergents formulations. Major areas of research and potential future applications of enzymes are also discussed in this part.

3.1 Basic thermodynamics and enzyme kinetics

Enzymes as biocatalysts catalyse many essential chemical reactions taking place in living systems. Reactions catalysed by enzymes proceed faster and at moderate temperatures and pH values (Atkins and Paula, 2001; Palmer, 1995). As with all other reactions, the basic laws of thermodynamics also apply to enzyme-catalysed processes. The first law of thermodynamics states that the change of energy caused by any event in a closed system is zero, i.e. energy cannot be created or destroyed, but it can be converted to other forms of energy to perform work. The second law states that the degree of entropy or degree of disorder is always increasing. Systems with high organisation and low entropy can be maintained by consumption of energy. Basically there are two forms of energy: one that can be used to perform work, also called free energy, and the other, which cannot. In any event, process or chemical reaction only happens spontaneously as a result of the decrease of the free energy, i.e. conversation of energy into work.

Gibbs defined the increase in free energy of a system, ΔG, as $\Delta G = \Delta H - T\Delta S$, where ΔH is the variation of enthalpy and ΔS is the variation of entropy for any event at a constant temperature T. For a process to take place in a spontaneous fashion, i.e. under thermodynamically irreversible conditions, ΔS must be higher than $\Delta H/T$, giving an overall increase in entropy of the system plus surroundings, as required by the second law of thermodynamics.

For any chemical reaction, the change of Gibbs free energy (ΔG) is the energy which is available to perform work as the reaction proceeds towards chemical equilibrium from the initial concentrations of reactants and products. If the sign of ΔG is negative, the system will release free energy to its surroundings as the reaction proceeds towards the equilibrium.

For the reaction:

$$\mathrm{R} \rightleftharpoons \mathrm{P} \qquad [3.1]$$

at a given temperature T, the free energy (ΔG) is given by:

$$\Delta G = -RT \cdot \ln \frac{[\mathrm{P_{eq}}]}{[\mathrm{R_{eq}}]} + RT \cdot \ln \frac{[\mathrm{P_0}]}{[\mathrm{R_0}]} \qquad [3.2]$$

where $[\mathrm{P_{eq}}]$, $[\mathrm{R_{eq}}]$ are the concentrations of the product and reactant at equilibrium, respectively and $[\mathrm{P_0}]$, $[\mathrm{R_0}]$ are the initial concentrations of the product and reactant, respectively. (To simplify the discussion we consider 'concentrations' instead of the more correct 'activities'.)

The equilibrium constant can be defined as $K_{eq} = \dfrac{[\mathrm{P_{eq}}]}{[\mathrm{R_{eq}}]}$ and

$$\Delta G = -RT \cdot \ln K_{\mathrm{eq}} + RT \cdot \ln \frac{[\mathrm{P_0}]}{[\mathrm{R_0}]} \qquad [3.3]$$

When the initial concentration of reactants and products is 1 M,

$$\Delta G = -RT \cdot \ln K_{eq} = \Delta G^{\theta} \quad [3.4]$$

where ΔG^{θ} is also called standard free energy change, i.e. the change of energy acquired from the initial concentrations of reactants and products of 1 M in reaching the equilibrium:

$$\Delta G = \Delta G^{\theta} + RT \ln \frac{[P_0]}{[R_0]} \quad [3.5]$$

Any reaction taking place will depend on the initial concentrations of products and reactants and also on the standard free energy. The balance between these two parameters will determine in which direction a reaction will run, provided that there is no interference from outside the system. However, in reality, this situation hardly ever exists and even with an overall negative free energy some reactions do not proceed. This can be explained by concepts such as collision theory and potential barrier or free activation energy. Chemical reaction can only occur when molecules collide. However, not all collisions are effective, i.e. not all colliding molecules will react with each other. This is mainly when colliding molecules do not have proper orientation or they do not have enough energy to react. This energy needed to initiate the reaction is called potential barrier or free activation energy.

Eyring postulated that every chemical reaction proceeds via the formation of an unstable intermediate between reactants and products, in the transition state. If the energy available in the system as collision energy is higher than a certain potential barrier, the reaction takes place. If not, the unstable intermediate returns to the initial state.

A catalyst accelerates a chemical reaction without changing its extent and with no overall thermodynamic effect, i.e. the amount of free energy change is the same in the presence or absence of the catalyst. The catalyst only reduces the amount of activation free energy resulting in a more stable transition state. In this fashion a more efficient transition intermediate is formed upon interaction between reactants and the catalyst (Fig. 3.1).

These principles can be applied to enzyme catalysis where an intermediate transition state is formed between a substrate and an enzyme accelerating the conversion of a substrate into a product. In this reaction, the substrate must fit precisely into the active site of the enzyme. Since enzymes are highly specific catalysts, it can be expected that the formation of the enzyme–substrate complex or the binding of the substrate in the active site will require only little energy. Consequently, enzymes are very effective catalysts, enhancing reactions up to 10000-fold more than the most effective chemical catalysts:

$$\underset{\text{(initial state)}}{E + S} \rightleftharpoons \underset{\text{(intermediate state)}}{ES} \rightarrow \underset{\text{(final state)}}{P} \quad [3.6]$$

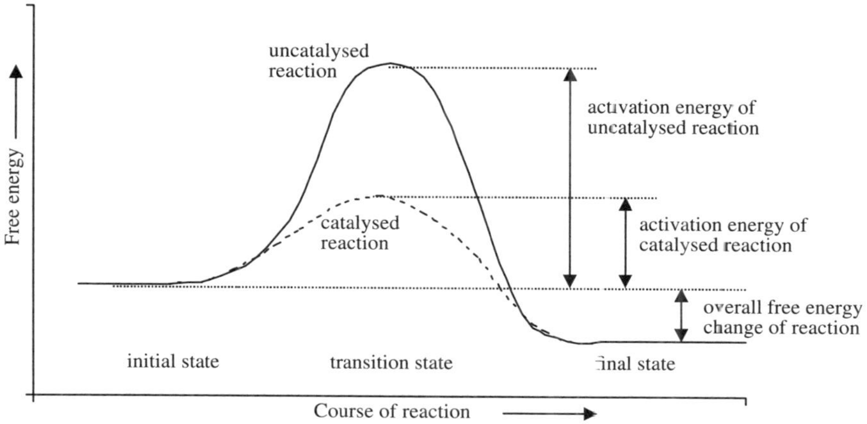

3.1 Change in free energy in catalysed and uncatalysed reactions.

3.2 Function of textile processing enzymes

Enzymes are proteins which are composed of folded peptide chains containing a wide range of amino acids. In living systems, mainly 20 different amino acids occur with a structural variety ranging from non-polar (aliphatic and aromatic) to acidic, basic and neutral polar properties. Therefore, depending on the amino acid composition and the three-dimensional (3D) structure of the protein, different microenvironments for catalysis exist at the active sites of enzymes. The high substrate specificity of enzymes is due to the individual architecture of the active site where only certain molecules can 'stereo-fit in'. Only little energy is required for the formation of this enzyme–substrate complex and therefore enzyme-catalysed reactions proceed very fast. Enzymes are generally active at mild temperatures because the enzyme proteins need to maintain their folded state in order to operate. Enzyme-catalysed reactions also proceed at mild pH-values; however, at the catalytic site, extreme acid or basic environments for catalysis can exist even when the reactions are carried out at neutral pH values. Depending on the organisms, some enzymes are also stable at extreme temperatures and pH values such as those from extremophiles living under these conditions.

Ionic bonds are important for the structural stability of a folded enzyme protein; it can be expected that the degree of protonation of the amino acid residues are a major issue and minor changes in pH have great effect on the stability of an enzyme and on its activity. Since the velocity of chemical reactions generally increases with temperature, the optimum temperature of an enzyme will be the highest temperature at which the enzyme protein can be maintained in a folded native state. Analysis of 3D structures and amino acid homology between thermophilic and non-thermophilic representatives of the same class indicate that enzyme structures with tighter loops, a higher level of glycosylation and/or higher level of crosslinkages have higher temperature optima and stability, and can also tolerate higher levels of agitation (Danson and Hough, 1998). In living systems enzymes catalyse essential chemical reactions under optimum reaction conditions via, for example, acid–base catalysis, covalent bonding or electron transfer mechanisms.

Acid–base catalysis is a common mechanism in enzyme reactions – for example, hydrolysis of ether, ester or peptide bonds, phosphate group reactions, additions to carbonyl groups and others. Acid catalysis usually involves donation of a proton by the catalyst while base catalysis involves abstraction of a proton. The side chains of the amino acids Asp, Glu, His, Cys, Tyr and Lys can be involved in general acid–base catalysis. Covalent catalysis involves rate enhancement by the transient formation of a covalent bond between the substrate and the catalyst, especially involving side chains of His, Cys, Asp, Lys and Ser.

Enzyme systems involving metal ion catalysis accelerate the reaction velocities by binding substrates in the proper orientation, mediating oxidation–reduction reactions and electrostatically stabilising or shielding negative charges. Metalloenzymes contain tightly bound metal ions such as Fe^{2+}, Fe^{3+}, Cu^{2+}, Zn^{2+} or Mn^{2+}, while metal-activated enzymes contain loosely bound metal ions such as Na^{+}, K^{+}, Mg^{2+} and Ca^{2+}. Electrostatic catalysis refers to the fact that when a substrate binds to an enzyme, water is usually excluded from the active site. This causes the local dielectric constant to be lower, which enhances charge–charge interactions at the active site. Proximity and orientation effects are also important in enzymatic reactions. The 3D structure of the enzyme can bring several reactive side chains into close proximity to the active site. Binding of the substrate at the active site can orientate the substrate for most efficient interaction with these side chains. Enzymes commonly used in textile processing will be discussed next.

3.2.1 Amylases

Amylases are widely used as desizing agents to remove starch from fabrics after weaving. Starch is a polysaccharide composed of glucose units primarily linked by α (1–4) glucosidic bonds with α (1–6) linked side chains. Depending on the number of branches, two types of polymer, amylose and amylopectin, are distinguished. Amylopectin accounts for around 70–80% of starch, containing branches at about every 20–24th glucose residue, while amylose is a much more linear polymer.

Enzymes involved in the complete degradation of starch are α-amylases, (EC 3.2.1.1), β-amylases (EC 3.2.1.2) and glucoamylase (EC 3.2.1.3).

α-Amylases hydrolyse randomly and are *endo*-acting on the α (1–4) bonds within the starch backbone, while the *exo*-acting glucoamylases cleave-off glucose units from the non-reducing ends of polysaccharides. The industrially less important β-amylases release only maltose units from the chains ends of starch polymer. However, these enzymes are not able to bypass branches. The enzyme mechanism usually involves two acidic amino acid residues, such as aspartic acid and/or glutamic acid along with a basic amino acid residue, e.g. histidine. In the same way as for most glycoside hydrolases there are two basic mechanisms: the α–α retaining mechanism of α-amylases or glucoamylase and the α–β inverting mechanism characteristic of β-amylases.

Since amylases have been naturally designed to act on an insoluble substrate, most amylases have an extra substrate binding domain. The substrate binding domain brings the catalytic domain into the close vicinity of the target substrate, enhancing the catalytic performance of the enzyme (Watanabe *et al.*, 2001 and Horvathova *et al.*, 2001). Studies on pancreatic amylases revealed that chlorine ions could also be essential for hydrolysis of starch. It is believed that chloride is required to increase the acidity at the active site thereby enhancing hydrolysis (Numao *et al.*, 2002).

Most commercial amylases used are crude mixtures of thermostable enzymes of bacterial origin. Amylases are activated by Ca^{2+} ions and it is known that these enzymes perform well in hard water rich in bivalent ions (Cavaco-Paulo, 1998). The presence of Ca^{2+} enhances the enzymatic reaction up to a certain level and is believed to stabilise the catalytic sites through structural organisation. The presence of calcium ions is a very important feature of bacterial thermostable amylases (stable up to 110°C) where they (up to three calcium cations) are believed to enhance the stability of enzyme by crosslinking the folded structure (Machius *et al.*, 1998).

α-Amylase activity is generally measured using starch as a substrate, monitoring the formation of reducing sugars using maltose as a standard (Numao *et al.*, 2002). Advanced activity measurement techniques for α- and β-amylases use oligosaccarides and follow the production of shorter chain sugars by chromatography (Watanabe *et al.*, 2001 and Horvathova *et al.*, 2001). β-amylase activity can be measured towards *p*-nitrophenyl-α-D-maltopentaose monitoring the release of *p*-nitrophenyl (Erkkilä *et al.*, 1998), while glucoamylase activity can also be determined following the release of *p*-nitrophenol from *p*-nitrophenyl-α-D-glucose (Lee *et al.*, 2001).

3.2.2 Cellulases

Cellulases are used in textile processing mainly for depilling and to obtain stone washing effects. Cellulases are also used as part of detergent formulations to enhance detergency, to improve brightness and to remove microfibrils (Cavaco-Paulo, 1998). In nature, cellulose, the world's most abundant polysaccharide, is enzymatically hydrolysed by the synergistic action of endo-β-1,4-glucanases (EC 3.2.1.4), cellobiohydrolases (EC 3.2.1.91) and β-glucosidases (EC 3.2.1.21). It has been suggested that endoglucanases (EGs) randomly cleave cellulose into smaller fragments generating new ends which are then hydrolysed endwise by the action of cellobiohydrolases. These latter enzymes are also thought to erode crystalline regions of cellulose making them more susceptible to EG attack (Wood, 1992). However, it is widely recognised that the classification of β-1,4-glucanases into exclusively *endo*- and *exo*-acting enzymes is, in many cases, not strictly definitive, as several enzymes have been isolated exhibiting both types of enzyme activities (Tomme *et al.*, 1996).

Like amylases, cellulolytic enzymes also employ an acid–base mechanism for the hydrolysis of their substrates, which involves two acidic amino acid residues such as aspartic acid and/or glutamic acid. However, in contrast to amylases, their catalytic activity and stability are generally independent of the presence of metallic ions. Hydrolysis of cellulose is catalysed via the β–β retaining mechanism or via the β–α inverting mechanism (Fig. 3.2).

Like amylases, cellulases have a catalytic domain and a substrate binding domain. In the past a considerable amount of work has been carried out to classify or group various cellulases and hemicellulases based on the degree of homology of the amino acid sequence of the various catalytic and binding domains of the enzymes. Several fungal cellulases have been grouped with enzyme families that are more closely related to known bacterial enzymes than they are to each other (Henrissat and Bairoch, 1996). Consequently, EGs from different microbial origins have shown similar substrate specificities, both on isolated and synthetic oligo- and polysaccharides and on their 'natural' substrates such as wood. In contrast, some closely related EGs have shown quite different substrate specificities (Tomme *et al.*, 1995). Important retaining cellulases belong to families 5, 7 and 12 while inverting cellulases are found in families 6, 8 and 45 (Davies and Henrissat, 1995; Henrissat, 1991; Henrissat and Bairoch, 1993, 1996). Cellulose binding domains (CBDs) of fungal origin are from family I (also called carbohydrate-binding module family 1) and they account for 33–36 amino acid residues, while bacterial cellulose binding domains from family II (also called carbohydrate-binding module family 2) are bigger with 105–120 amino acids. Family I CBDs are present in almost all cellulase preparations commonly used in textile and detergent applications and they bind cellulosic fibres reversibly while family II CBDs bind cellulose more strongly (Cavaco-Paulo *et al.*, 1999).

Cellulase activities can be measured towards insoluble cellulose in the form of filter paper, or microcrystalline cellulose eventually swollen in

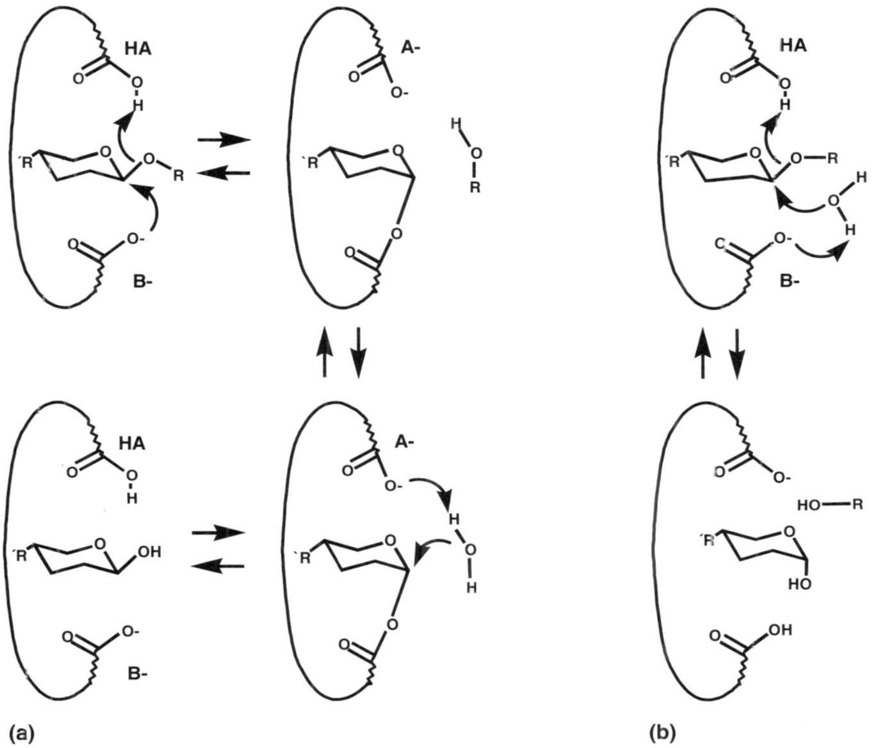

3.2 (a) β–β Configuration retaining mechanism of cellulose hydrolysis by cellulase enzymes. (b) β–α Configuration inversion mechanism of cellulose hydrolysis by cellulase enzymes.

phosphoric acid. Reducing sugars released can be monitored by, for example, the DNS (dinitro salicylic acid) method (Ghose, 1987). In a commercial mixture, the values obtained with this method reveal the hydrolysis rate caused by the synergistic action of EG and cellobiohydrolase activities. EG activity can be measured towards carboxymethylcellulose (CMC) following the release of reducing sugars or the decrease of viscosity of CMC solutions. Enzyme preparations completely free of EG activity do not show any activity towards CMC solutions (Cavaco-Paulo *et al.*, 1996 and Ghose, 1987).

3.2.3 Pectinolytic enzymes

Pectin-degrading enzymes have received much interest for their use in the pretreatment of textile fabrics ('bioscouring') prior to dyeing. The removal of pectin components from the cotton cell wall is claimed to improve fibre hydrophilicity, to facilitate dye penetration and to contribute to substantial water savings when compared to the traditional alkaline scouring process.

In nature, three major classes of enzymes are involved in the degradation of pectins: pectin esterases, polygalacturonases and pectin lyases. Pectin esterases (EC 3.1.1.11) catalyse the de-esterification of polymethylgalacturonate forming pectic acid (polygalacturonate). Pectin esterases commonly employ a Ser-His-Asp catalytic triad such as acetylxylan esterases to catalyse deacetylation, but other mechanisms such as a Zn^{2+} catalysed deacetylation may also be considered for some families (Fig. 3.3). Polygalacturonases cleave α(1–4) glycosidic linkages in polygalacturonate and can be divided into two groups according to their mode of action on the polymer: endopolygalacturonases (EC 3.2.1.15) hydrolyse randomly within pectic acid while exopolygalacturonases (EC 3.2.1.67) cleave in a sequential fashion generally from the non-reducing end of the pectin chain. Only little is known about the stereochemistry of the hydrolysis reaction but there seem to exist both retaining and inverting endo- and exopolygalacturonases (Biely *et al.*, 1996).

Pectin lyases cleave polygalacturonate or pectin chains via a β-elimination resulting in the formation of a double bond between C4 and C5 at the non-reducing end (Fig. 3.4). There are three major types of lyases: endopolygalacturonate lyases (EC 4.2.2.2) which randomly cleave polygalacturonate chains, exopolygalacturonate lyases (EC 4.2.2.9) which cleave at the

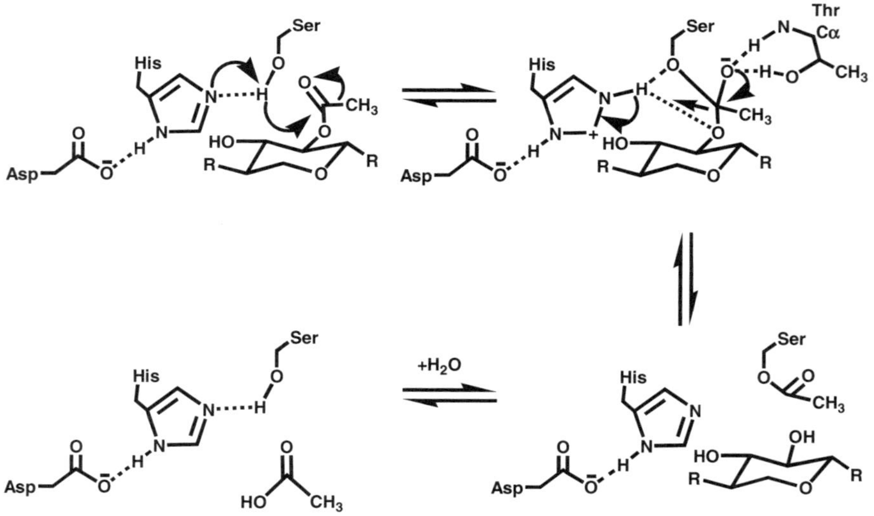

3.3 Mechanism of hydrolysis of the acetyl xylan esterase by the triad Asp-Hist-Serine (Hakulinen *et al.*, 2000).

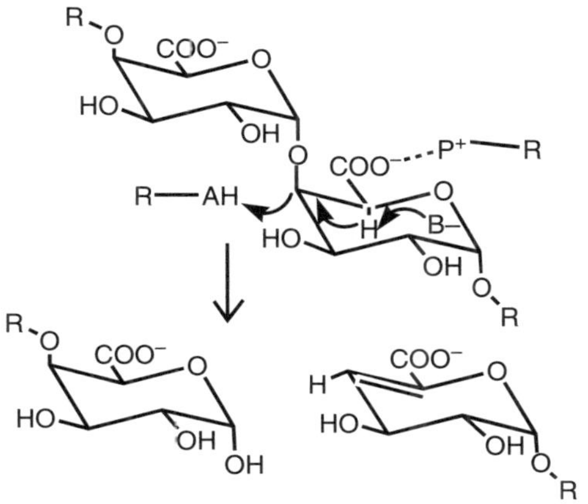

3.4 Schematic diagram of α-1,4-polygalacturonic acid cleavage by the β-elimination mechanism (Herron *et al.*, 2000).

chain end of polygalacturonate yielding unsaturated galacturonic acid and endopolymethylgalacturonate lyases (EC 4.2.2.10) which randomly cleave pectin (Sakai *et al.*, 1993 and Whitaker, 1989).

Enzymes that solubilise pectin from protopectin are called protopectinases (Sakamoto and Sakai, 1994). Various enzymes including endoarabinases, pectate lyases and polygalacturonases can show this ability and are called protopectinases to distinguish these enzymes from classical endoarabinases, pectate lyases and polygalacturonases without pectin-releasing activity (Ferreyra *et al.*, 2002; Matsumoto *et al.*, 2000).

Pectin esterase activity can be measured with pectin (polymethylgalacturonate) as a substrate monitoring the pH change in the solution caused by the formation of carboxylic acids. Another possibility for detecting pectin esterase activity is to measure methanol released from the substrate (Sakai *et al.*, 1993). It should be noted that at pH values higher than 7, pectin can be hydrolysed. The determination of endo- and exopolygalacturonase activity towards pectin can be followed by the formation of reducing sugars. Since the enzyme catalyses the depolymerisation of a soluble substrate, an alternative assay method measures the decrease in viscosity. To distinguish between *endo* and *exo* enzymes, chromatographic techniques that identify short chain oligosaccharides formed or viscosity methods can be used (Whitaker, 1989). Pectin lyase activity towards pectic acid can be measured by monitoring the absorbance change at 235 nm caused by the formation of the unsaturated product. Another method for determining pectin lyase activity is based on the formation of a red complex caused by the reaction between the unsaturated galacturonic acid and thiobarbituric acid

(Whitaker, 1989). Protopectinase activity is measured using protopectin as substrate (Sakai *et al.*, 1993).

3.2.4 Esterases

Esterases have been suggested as useful components of detergent formulations to remove lipid-based stains from textiles while some esterases have been claimed to hydrolyse polyester. Esterases hydrolyse ester bonds and their classification is based on the type of ester bond hydrolysed.

Esterases with applications in textile processing include carboxylesterases (EC 3.1.1.1) which hydrolyse carboxylic esters yielding the corresponding alcohol and carboxyl anion, arylesterases (EC 3.1.1.2) which hydrolyse phenyl acetate to phenol and acetate and triacylglycerolesterases (EC 3.1.1.3) which hydrolyse triacylglycerol giving a diacyl glycerol and fatty acid anion. The latter enzymes are better known as lipases. Many esterases are multifunctional enzymes and they can work as carboxylesterases, lipases and others. In the hydrolysis reaction, the catalytic triad Ser-His-Asp can be involved in the same way as some proteases (Fig. 3.3). Esterases can show activation in water/lipid interfaces which has been described particularly for lipases dependent on the pH of the medium (Petersen *et al.*, 2001 and Cambillau *et al.*, 1996). Cutinases, a class of acyl esterases, are particularly active on cutin and do not show any interfacial activity even though these enzymes have been described as hydrolysing triglycerides (Cambillau *et al.*, 1996).

Carboxylesterase activity is measured, for example towards *o*-nitrophenyl butyrate; arylesterases are measured, for example towards phenyl acetate; and lipases are assayed towards triacylglycerols (e.g. olive oil). These reactions can be monitored via pH change or alternatively via numerous colorimetric methods following the products formed (Bergmeyer, 1974).

3.2.5 Proteases

Proteases are important components of detergent formulations for removing protein stains (egg, blood etc.) from textiles. Additionally, proteases have a useful potential in silk and wool processing. Proteases or, more correctly, peptidases hydrolyse peptide bonds in soluble and insoluble peptides and form the group EC 3.4.X.X. of hydrolases. Peptidases can be divided into endopeptidases and exopeptidases, which cleave peptide bonds within the protein or release amino acids sequentially from either the N- or C-terminus, respectively. Proteases have been grouped into families and clans according to the homology in their catalytic domains. According to the mechanism of hydrolysis these enzymes have been grouped into serine, cysteine, aspartic and metallo-proteases. Representatives of serine proteases are mammalian chymotrypsin and trypsin or the bacterial subtilisin with the catalytic triad consisting of Ser-His-Asp. Cysteine-type proteases include papain with the catalytic triad of Cys-His-Asn while in the catalytic reaction of aspartic type proteases such as pepsin two aspartates are involved. Metallopeptidases such as thermolysin generally contain a Zn atom which is involved in the catalytic reaction.

Protease activities can be measured towards proteins such as casein or haemoglobin by following the release of hydrolysis products colorimetrically. Other more specific substrates are used if the hydrolysis of a certain peptide bond is targeted (Beynon and Bond, 1996; Bergmeyer, 1974).

3.2.6 Nitrile-hydrolysing enzymes

Nitrilases have been shown to improve dye uptake and hydrophilicity of acrylic fibres. These improved properties are achieved by enzymatic conversion of nitrile groups into carboxylic acid groups at the surface of acrylic fibres.

In nature, three different groups of enzymes are involved in the microbial hydrolysis of nitriles (Fig. 3.5). Nitrilases (EC 3.5.5.1, EC 3.5.5.7) hydrolyse nitriles to the corresponding carboxylic acids forming ammonia; nitrile hydratases (EC 4.2.1.84) form amides from nitriles which can subsequently be hydrolysed by amidases (EC 3.5.1.4) (Tauber *et al.*, 2000). Formerly, nitrilases were thought to hydrolyse exclusively aromatic substances while aliphatic nitriles were believed to be degraded by a nitrile hydratase/amidase enzyme system. However, recent investigations have shown that this strict rule does not always apply.

The reaction mechanism, regulation and photoactivation of nitrile hydratases, which usually consist of α- and β-sub-units containing either non-heme iron or cobalt atoms have been studied in detail (Kobayashi and Shimizu, 1998). Hydrolysis of nitriles and amides by nitrilases and amidases,

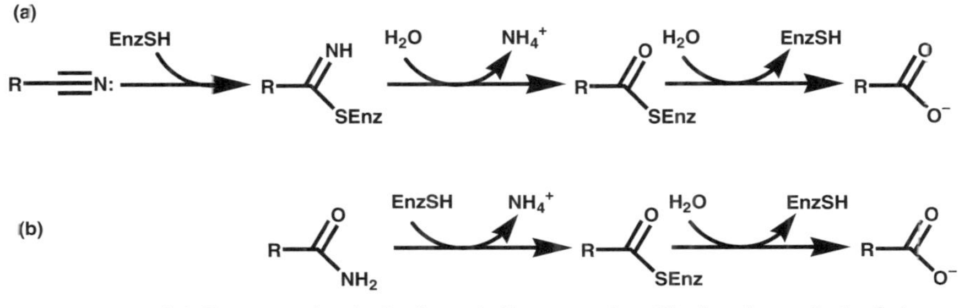

3.5 (a) Enzymes hydrolysing nitriles are classified to branch 1 of the nitrilase superfamily. (b) The amidase reaction is the most frequently observed activity of enzymes classified to other branches of the nitrilase superfamily.

respectively, is catalysed via a thiol acyl enzyme intermediate involving the Glu-Lys-Cys catalytic triad (Pace and Brenner, 2001). Nitrile-hydrolysing enzymes can be assayed towards nitriles such as to acetonitrile and/or benzeno nitrile methods yielding the respective amides and carboxylic acids and the reaction is generally followed by chromatography (Tauber *et al.*, 2000).

3.2.7 Catalases, peroxidases and catalase-peroxidases

Catalases (EC 1.11.1.6) can be used in textile processing for the removal of residual hydrogen peroxide after bleaching while peroxidases (EC 1.11.1.7) have a potential for dye decolourisation after dyeing (Gudelj *et al.*, 2001).

Catalases convert hydrogen peroxide into water and oxygen showing first order kinetics. This loop reaction starts by oxidation of the catalase to compound I by one molecule of hydrogen peroxide yielding water and regeneration via production of oxygen from the second molecule of H_2O_2 (see reactions [3.7] and [3.8]). Usually catalases have heme-containing prosthetic groups. Bifunctional catalase-peroxidases can oxidise substrates other than H_2O_2 (Zamocky *et al.*, 2001). In the first step catalase-peroxidase compound I is formed because of oxidation by peroxide. Compound I is situated two oxidation equivalents higher and has a porphyrin-π-cation radical with an iron (IV) centre and can be reduced to the starting form by hydrogen peroxide. Alternatively compound I can be reduced by a one-electron reduction to Compound II, which is the peroxidase reaction. Compound II has an amino acid radical ($R^{\bullet}$) and iron (III). Finally, Compound II is reduced to the starting form by a second one-electron reduction.

$$\underset{\text{Ferric enzyme}}{\mathrm{Fe(III)}\ldots\mathrm{R}} + \mathrm{H_2O_2} \rightarrow \underset{\text{Compound I}}{[\mathrm{Fe(IV)}{=}\mathrm{O}\ldots\mathrm{R}]^{\bullet+}} + \mathrm{H_2O} \qquad [3.7]$$

$$\underset{\text{Compound I}}{[\mathrm{Fe(IV)}{=}\mathrm{O}\ldots\mathrm{R}]^{\bullet+}} + \mathrm{H_2O_2} \rightarrow \underset{\text{Ferric enzyme}}{\mathrm{Fe(III)}\ldots\mathrm{R}} + \mathrm{O_2} + \mathrm{H_2O} \qquad [3.8]$$

$$\underset{\text{Compound I}}{[\mathrm{Fe(IV)}{=}\mathrm{O}\ldots\mathrm{R}]^{\bullet+}} + \mathrm{AH_2} \rightarrow \underset{\text{Compound II}}{[\mathrm{Fe(III)}{=}\mathrm{O}\ldots\mathrm{R}]^{\bullet+}} + \mathrm{AH}^{\bullet} \qquad [3.9]$$

$$\underset{\text{Compound II}}{[\mathrm{Fe(III)}{=}\mathrm{O}\ldots\mathrm{R}]^{\bullet+}} + \mathrm{AH_2} \rightarrow \underset{\text{Ferric enzyme}}{\mathrm{Fe(III)}\ldots\mathrm{R}} + \mathrm{AH}^{\bullet} \qquad [3.10]$$

Catalases catalyse reactions [3.7] and [3.8] and catalase-peroxidases catalyse reactions [3.7], [3.9] and [3.10] (Zamocky *et al.*, 2001).

During lignin degradation, fungi employ so-called manganese-peroxidases (EC 1.11.1.13) requiring the presence of manganese ions:

$$2\mathrm{Mn(II)} + 2\mathrm{H^+} + \mathrm{H_2O_2} \rightarrow 2\mathrm{Mn(III)} + 2\mathrm{H_2O} \qquad [3.11]$$

These enzymes and other peroxidases can also be used for textiles dye degradation.

Catalase and peroxidase activities can be measured spectrophotometrically following the degradation of hydrogen peroxide at 240 nm and the colour change during the oxidation of various substrates, respectively (Gudelj *et al.*, 2001).

3.2.8 Laccases

Laccases in combination with redox mediators are used in textile processing to bleach denim fabrics, decolourising indigo. Research efforts have been made to use laccase as a bleaching and/or oxidative coupling agent for dyeing animal fibres and human hair.

Laccases (1.10.3.2) are unspecific oxidoreductases which catalyse the removal of a hydrogen atom from the hydroxyl group of *ortho* and *para*-substituted mono- and polyphenolic substrates and from aromatic amines by one-electron abstraction while the cosubstrate oxygen is reduced yielding water. Free radicals formed in this reaction from the substrates are capable of undergoing further depolymerisation, repolymerisation, demethylation or quinone formation. The rather broad substrate specificity of laccases may be additionally expanded by addition of redox mediators such as ABTS [2,2′-azinobis(3-ethylbenzthiazoline-6-sulphonate)]. These blue oxidases typically contain four copper atoms per polypeptide chain distributed in three different copper binding sites (types I, II and III). It is believed that the initial oxidation of the enzyme by oxygen occurs at the T2/T3 site followed by an electron transfer from T1 to T2/T3 site and further oxidation of the substrate (Gianfreda *et al.*, 1999). Laccases are assayed following the oxidation of various substrates such as dimethoxyphenol, ABTS or syringaldizine spectrophotometrically (Abadulla *et al.*, 2000).

3.3 Homogeneous and heterogeneous enzyme catalysis and kinetics

3.3.1 Enzyme kinetics of homogenous systems

Kinetics is the study of reaction rates measured by the change in quantity of reactants with time. Chemical kinetics is ruled by the law of mass action. This law states that the rate of reaction is proportional to the product of the activities of the reactants (A,B) considering the stoichiometric constants (*a*,*b*) of each reactant:

$$aA + bB \rightarrow \text{production} \qquad [3.12]$$

$$v = k[A]^a \cdot [B]^b \qquad [3.13]$$

For practical proposes activity can be replaced by concentration measured in molarity. The order of the reaction is a for the reactant A and b for the reactant B and of general order $a + b$. The rate v of the reaction:

$$A \rightarrow P \qquad [3.14]$$

can be described as:

$$v = -\frac{d[A]}{dt} = +\frac{d[P]}{dt} = k[A] \qquad [3.15]$$

where k is a rate constant and [A] and [P] are the concentrations of reactant A and the product P at the time t.

$-\frac{d[A]}{dt}$ and $+\frac{d[P]}{dt}$ describe the rate of decrease of A and increase of P, respectively. The rate of reaction at various times can be found by taking tangents in a plot of concentration change versus time and calculating their gradients. The reaction orders for each reactant are experimentally determined by measuring the initial reaction rates at different initial concentrations of this reactant.

These rules can be also applied to enzymatic reactions. Enzyme-catalysed reactions occurring in homogenous media where both the substrate(s) and the enzyme are in solution show a general trend: the initial rates are first order at low substrate concentrations and zero order at very high substrate concentrations (Fig. 3.6).

This behaviour can be explained by the formation of an enzyme–substrate complex:

$$\underset{\text{(initial state)}}{E + S} \underset{k_b}{\overset{k_a}{\rightleftharpoons}} \underset{\text{(intermediate state)}}{ES} \overset{k_c}{\rightarrow} \underset{\text{(final state)}}{E + P} \qquad [3.16]$$

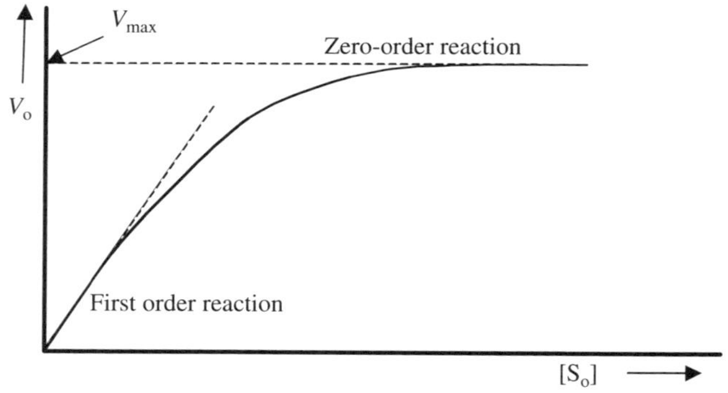

3.6 Typical initial rates – substrate dependence.

During the reaction all the enzyme is usually present in the form of the enzyme–substrate complex ES if the concentration of the enzyme is much lower than the concentration of the substrate. A quasi-steady state for the enzyme–substrate complex can be assumed:

$$-\frac{\mathrm{d[ES]}}{\mathrm{d}t} = k_a[\mathrm{E}][\mathrm{S}] - k_c[\mathrm{ES}] - k_b[\mathrm{ES}] = 0 \qquad [3.17]$$

Using the mass balance for the enzyme in free or associated form:

$$[\mathrm{E}] = [\mathrm{E_0}] - [\mathrm{ES}] \qquad [3.18]$$

From the equation:

$$k_a[\mathrm{E}][\mathrm{S}] = (k_c + k_b)[\mathrm{ES}] \qquad [3.19]$$

the concentrations of ES can be determined to give:

$$k_a[\mathrm{E_0}][\mathrm{S}] - k_a[\mathrm{ES}][\mathrm{S}] = (k_c + k_b)[\mathrm{ES}] \qquad [3.20]$$

or

$$[\mathrm{ES}] = \frac{[\mathrm{E_0}][\mathrm{S}]}{k_m + [\mathrm{S}]} \qquad [3.21]$$

where:

$$k_m = \frac{k_b + k_c}{k_a} \qquad [3.22]$$

With the reaction rate for the dissociation of the enzyme–substrate complex and formation of the product:

$$v = k_c[\mathrm{ES}] \qquad [3.23]$$

the result is the Michaelis–Menten equation:

$$v = \frac{k_c[\mathrm{E_0}][\mathrm{S}]}{k_m + [\mathrm{S}]} \qquad [3.24]$$

For the maximum reaction rate:

$$v_{\max} = k_c[E_0] \qquad [3.25]$$

we obtain:

$$v = \frac{v_{\max}[\mathrm{S}]}{k_m + [\mathrm{S}]} \qquad [3.26]$$

k_m gives the substrate concentration $[S_0]$ at v_0 as 1/2 $v_{\max}$. k_m is also called the Michaelis–Menten constant.

For an enzymatic reaction, k_c is also called the turnover number k_{cat}, which represents the maximum number of substrate molecules that can be

Table 3.1 Classical linearisation methods for Michaelis–Menten equation

Equation	Plot
$\frac{1}{v_0} = \frac{k_m}{v_{max}} \frac{1}{[S_0]} + \frac{1}{v_{max}}$	Lineweaver–Burk
$v_0 = -k_m \frac{v_0}{[S_0]} + v_{max}$	Eadie–Hofstee
$\frac{[S_0]}{v_0} = \frac{1}{v_{max}}[S_0] + \frac{k_m}{v_{max}}$	Hanes–Woolf

converted by a unit of time. In more complex enzymatic reactions involving several steps and various intermediates following Michealis–Menten kinetics k_{cat} can be seen as a function of several individual reaction rates.

k_{cat}/k_m can be regarded as the catalytic efficiency of an enzyme. Comparing the values of k_{cat}/k_m for different substrates and one enzyme, this value can be regarded as the specificity of an enzyme towards a substrate. k_m can be regarded as the affinity of an enzyme towards a substrate, or the stability of enzyme substrate complex, i.e. higher k_m, lower affinity and lower stability.

The determination of v_{max} and k_m may involve the determination of initial reaction rates for several substrate concentrations at a given enzyme concentration. The classical linearisation methods of the Michealis–Menten equation have been employed over the years to determine the parameters while some of them give considerable errors (Table 3.1). Nowadays Michealis–Menten parameters can be estimated directly by non-linear regression methods using computer programs.

Although many enzyme-catalysed reactions can be described by this simple Michealis–Menten model, some enzymes like catalases show very high turnover numbers where the enzyme can never be saturated with its substrate hydrogen peroxide because, for example, H_2O_2 destroys the catalase at very high concentrations. On the other hand, a lot of biological reactions involve more than one substrate and complex enzyme systems. In several reactions involving more than one substrate, the Michealis–Menten model can still be applied to one individual substrate provided that the other substrates are present in excess. However, no information can be obtained about the exact multisubstrate reaction mechanism (e.g. random, ordered and ping-pong).

3.3.2 Enzyme catalysis in heterogeneous systems

In heterogeneous systems at least the catalyst or one of the reactants or products is present in a different phase from the others. An example of the application of an insoluble enzyme used to convert soluble substrates related to textile processing is the application of immobilised enzymes such as laccases or catalases for the treatment of dyeing and bleaching effluents, respectively. Most of the enzyme applications in textile processing, however, involve heterogeneous systems consisting of soluble enzymes and insoluble substrates in the form of textile materials or their components. Classical examples of heterogeneous enzymatic catalysis are the enzymatic hydrolysis of insoluble polymers like wool or silk by proteases and cotton or synthetic fibres by cellulases. Most carbohydrolases such as cellulases, pectinases and amylases are known to have substrate binding domains. These enzymes have been designed by nature with a special peptide binding to the substrate which is the driving force of the soluble enzyme in attacking an insoluble substrate. It is believed that substrate binding domains increase the concentration of the enzyme nearby the substrate and that they are essential for efficient enzymatic hydrolysis of insoluble polymers.

Often there is a limitation in terms of accessibility of the insoluble substrate to the soluble enzyme. The enzyme can only access the outer parts of the substrate at the liquid–solid interface while inner parts are only accessible when the outer parts are removed. Interestingly, the synergistic action between several cellulase components during hydrolysis of crystalline cellulose has only been observed at lower concentrations, i.e. when there was no competition between the different cellulase components for the hydrolytic substrate sites (Woodward *et al.*, 1988). These facts are of particular importance for industrial applications using solubles enzyme for the modification of insoluble substrates, since sometimes very high enzyme concentrations are used. It is most likely that in soluble enzyme–insoluble substrate systems, enzymes are saturating the few available substrate sites. It is obvious that classical Michealis–Menten kinetics cannot be applied to these systems because of the simple fact that a solid concentration cannot be determined. In Michealis–Menten kinetics, saturation of the enzyme by the substrate is verified, but in soluble enzyme–insoluble substrate systems it is the substrate that is saturated by enzyme; therefore a relationship has been suggested (Bailey, 1989) between v_0 and $[E_0]$:

$$v_0 = \frac{v_{max}[E_0]}{k_e + [E_0]} \qquad [3.27]$$

In a similar fashion to classical kinetics, interchanging S_0 by E_0, the parameters would have similar significance but could not be interchanged, since enzyme concentration could barely be expressed in molar units. Empirically these expressions have been verified and the estimated parameters are of prime importance to characterise different enzymes systems and different process conditions (Cavaco-Paulo *et al.*, 1998).

The former dependence of initial enzyme rate on enzyme concentration can be extended to all conversion times and since the performance of an enzyme is directly dependent on reaction rates, a performance (P) benefit can also be measured as a function of enzyme dosage (D_e).

$$P = \frac{P_{max} D_e}{D_{e,0.5P_{max}} + D_e} \qquad [3.28]$$

P_{max} is the maximal performance and $D_{e,0.5P_{max}}$ is the enzyme dosage for half of maximal performance. This is of prime importance for optimisation of industrial enzyme treatments of soluble enzyme–insoluble substrate systems (Ee *et al.*, 1997).

Turnover numbers for soluble substrates are usually much higher than for insoluble substrates. For cellulases from *Humicula insolens*, it is known that turnover numbers for soluble substrates such as carboxymethylcellulose are 20 times higher than for insoluble substrates such as acid-swollen cellulose (Schulein, 1997). These low turnover numbers might allow the interaction of enzyme and the substrate almost in a quasi-reversible fashion. The number of enzyme sites on the insoluble substrate surface can be determined using typical surface adsorption isotherms, such as the monolayer Langmuir type model (Cavaco-Paulo, 1998):

$$\frac{E_{ads}}{E_{max}} = \frac{KC_e}{1 + C_e} \qquad [3.29]$$

where E_{ads} is the amount of adsorbed enzyme per substrate mass, E_{max} is the maximum amount of adsorbed enzyme, K is the adsorption constant of the enzyme on the substrate and C_e is the free enzyme concentration in solution. Comparative values of E_{ads} and K can explain important characteristics about the individual enzyme substrate interaction (Cavaco-Paulo, 1998).

The adsorption of enzymes at the surface of an insoluble substrate only follows the Langmuir isotherm law when a monolayer of enzymes is formed. This is not the case when enzymes agglomerate on the substrate surface or the enzyme penetrates into a porous substrate.

Kinetic models for immobilised enzymes strongly depend on the immobilisation method. Enzymes can be attached to solid materials (glass, alumina, synthetic and natural polymers) via a range of different approaches from entrapment to covalent linking. Provided that the carrier material does not influence diffusion of the reactants and the enzyme, kinetic models for soluble enzymes can be used. However, the nature of the carrier material can lead to higher or lower concentrations of the substrate in proximity to the immobilised enzyme. Interaction of charged carrier materials and charged substrates lead to changes in the k_m values and can be described by models based on the Maxwell–Boltzmann distribution of the charged substrate between the polyelectrolyte phase and the solution.

On the other hand, the accessibility of the enzyme active site can be decreased.

The reaction rate for the immobilised enzyme can be described by extension of the Michaelis–Menten model with an efficiency factor η (Bisswanger, 2002):

$$v' = \eta v = \eta \frac{v_{\max}[S]}{k_m + [S]} \qquad [3.30]$$

The factor η is dependent on the substrate concentration and for $\eta = 1$ the reaction obeys the Michaelis–Menten model for the soluble enzyme while for lower values the reaction is predominantly diffusion controlled. A number of models have been developed to describe enzyme reactions controlled by external diffusion phenomena on the enzyme carrier layer and internal diffusion within porous carrier materials to the enzyme (Bisswanger, 2002). Kinetic models for immobilised enzymes usually do not consider changes in the enzyme itself which are especially likely during covalent modification.

3.3.3 Enzyme activity

We have learned in the previous section that enzymes are specific to a limited number of substrates. Especially for dosing enzymes in industrial applications, it is very important to know the exact activity of enzymes in commercial preparations which usually cannot be deduced from protein concentrations or other parameters. Assays for the determination of the activity of a certain enzyme are standardised and use well-defined substrates and reaction conditions (pH, temperature etc.). Enzyme activity is expressed as katals where one katal (kat) is defined as the amount of enzyme transforming one mole of substrate per second under standard conditions of temperature, optimal pH and optimal substrate concentration ($v_{\max}$). Previously, enzyme activity was expressed as International Units (IU) corresponding to the transformation of 1 micromole of substrate per minute ($1\,\text{IU} = \mu\text{mol}\,\text{min}^{-1} \approx 16.67\,\text{nkat}$):

$$-\frac{d[S]}{dt} = v_{\max} = K[E] \qquad [3.31]$$

3.3.4 Enzyme inhibition or enhancement

Inhibition or enhancement of the rate of enzyme-catalysed reactions involves specific interaction of agent (inhibitors or enhancers) with catalytic or regulatory sites on the enzyme or the enzyme substrate intermediate. There are three types of reversible inhibition including competitive, non-competitive and uncompetitive inhibition. Competitive inhibitors usually have a structure similar to the substrate and they bind in competition with the actual substrate at the substrate binding site without being transformed. At high substrate concentrations v_{max} remains unchanged while higher k_m values result. A non-competitive inhibitor does not influence the binding of the substrate but it prevents the enzyme–substrate complex from dissociating. In this case v_{max} is reduced while k_m remains the same. The non-competitive inhibitor can bind both to the free enzyme and to the enzyme–substrate complex while so-called uncompetitive inhibitor can only react with the enzyme–substrate complex changing both k_m and v_{max}. The rate equations for the different types of inhibition based on dissociation constants k_I of the enzyme (E)–inhibitor (I) complexes are presented in equations [3.32] to [3.34]:

$$\text{competitive} \quad v_0 = \frac{v_{max} \cdot [S_0]}{k_m\left(1+\frac{[I]}{k_i}\right)+[S_0]} \qquad [3.32]$$

$$\text{non-competitive} \quad v_0 = \frac{v_{max} \cdot [S_0]}{\left(1+\frac{[I]}{k_i}\right)(k_m+[S_0])} \qquad [3.33]$$

$$\text{uncompetitive} \quad v_0 = \frac{v_{max} \cdot [S_0]}{k_m+[S_0]\left(1+\frac{[I]}{k_i}\right)} \qquad [3.34]$$

Both strong non-covalent binding (binding constants of $>10^{-10}$ M) and covalent binding of the inhibitor to the enzyme can lead to irreversible inhibition. A time-dependent decrease of the enzyme activity is characteristic of irreversible inhibition.

3.3.5 Stability of enzymes and half-life times

For the industrial application of enzymes both the stability of the enzymes in the process and during storage is of great interest. At extreme pH values drastic changes in the charge on the enzyme molecule can cause irreversible destruction of the native protein structure. Usually this so-called denaturation shows a first order exponential decrease in the enzyme activity. Similarly, the native structure of enzymes can be destroyed at high temperatures, by detergents and other substances. In textile applications particularly, a number of auxiliaries could potentially interact with enzymes. As an example it has been shown that sequestering agents can affect the activity of laccase chelating copper which is essential for the enzyme function. Also, multimeric enzymes like catalases are deactivated by surfactants separating the individual units. These important issues are discussed in more detail in Chapter 5 where techniques for the stabilisation of enzymes are also presented. Usually enzyme stabilities are described as half-life times of the enzyme activity. The deactivation of an enzyme generally follows a first order reaction. Based on the first order deactivation of the enzyme

$$-\frac{d[E]}{dt} = k_d[E] \qquad [3.35]$$

where k_d is the deactivation constant. After integration

$$\ln\frac{[E_0]}{[E]} = k_d \cdot t \qquad [3.36]$$

The half-life time $t_{1/2}$ is defined as:

$$t_{1/2} = \frac{\ln 2}{k_d} \qquad [3.37]$$

3.4 Major enzymatic applications in textile wet processing

Enzymes can be applied in several steps of textile wet processing and in formulation of detergent powders. Since the major textile finishing process is coloration, classical finishing processes can be divided into preparation for coloration and after-coloration steps. Coloration might be done during fibre extrusion of synthetic fibres, on a bundle of fibres, on yarns, on fabrics or on garments. The sequence of processes depends on the demands of the market for the characteristics of a final product but depends essentially at which stage the coloration process is done. (If fashion or market regulations demand materials in the raw state, they are supplied unfinished.) Preparation for coloration steps generally involves the removal of impurities, natural coloured pigments, sizes and lubricants. Preparation of synthetic fibres also involves thermal treatments for uniform dyeing. After coloration, processes include chemical and mechanical processes. Industrial laundering and home washing of garments can be also included in the after-coloration processes. To give an overview of enzymatic applications in textiles a brief characterisation of major wet processing steps before and after coloration and during coloration itself will be presented.

3.4.1 Overview of traditional wet processing

Enzymes can be applied in several steps of textile wet processing and in formulations of detergent powders.

3.4.1.1 Preparation for coloration

Preparation for coloration processes aims to prepare the textile materials to receive dyes or pigments with high fastness properties. In preparation, all impurities and natural colored pigments have to be removed. Generally preparation for coloration is similar for all colours, but is more stringent for whites and lighter shades. Major processes during preparation are singeing, desizing, scouring, washing-off, bleaching, mercerising, carbonisation and thermal treatments.

Singeing consists of treatment with flames to burn out fuzz fibres directly from fabrics and is applied mainly on cellulosic materials and their mixtures. Desizing is the removal of sizes that are added to yarns to prevent breaks and stops during the weaving process. Desizing is only done on woven fabrics. Depending on the chemical nature of the size, removal could be effected by hydrolysis or oxidative processes or both. Scouring is the removal of natural impurities of natural fibres and can be applied to fibres, yarns, knitted or woven fabrics and garments. Scouring is done by neutral or alkaline washing with detergents. Washing-off is the removal of lubricants added during the spinning, knitting or weaving process to reduce friction and electrostatic energy. Washing-off is also done with detergents. Both the scouring and washing-off processes improve the hydrophilicity of the textile material and help the dyes to penetrate the fibres. Scouring is usually applied to natural fibres and washing-off is usually applied to synthetic fibres.

Carbonisation is a process applied to wool fibres to remove the vegetal soils, by treatment with sulfuric acid. Digested cellulosic impurity residues are removed from the fibres by brushing and suction. Bleaching is the removal of naturally coloured pigments in natural fibres. Nowadays it is done with hydrogen peroxide in alkaline conditions and it applied to fibres, yarns, fabrics or garments. Bleaching treatments are performed in more gentle alkaline conditions on wool and in very caustic conditions in linen. Bleaching of bast fibres most of the time involves a double bleaching process to achieve good whiteness results. Bleaching can be combined with scouring for cellulosic knitted fabrics and combined with desizing and scouring for cellulosic wovens using more concentrated alkaline conditions where sizes and natural impurities are removed along with natural pigments.

Mercerisation is the treatment of cellulosic fibres with highly concentrated solutions of caustic soda (300 g/L) under tension. Mercerisation induces intercrystalline swelling of cellulose, changing the crystal structure of cellulose I to mixture of cellulose I and II, the changes in the microstructure of cellulose being responsible for improved properties such as fibre strength, dye uptake brightness and hydrophilicity.

Thermal treatment, also called thermosetting, is used for all synthetic fibres with the aim of giving the same thermal history to the textile in order to achieve even results in further dyeing.

3.4.1.2 Coloration

Coloration is a major process in textile finishing and consists of the fixation of dyes and pigments in textile materials with high fastness properties. There are several classes of dyes, depending on the process of application and on the chemical nature of the fibre. Major classes of dyes for cellulosic fibres are direct, vat and reactive dyes. Major classes for protein fibres are acid and reactive dyes. Disperse dyes can be used mainly for polyester fibres, cationic dyes for acrylics and acid or disperse dyes for polyamides.

All dyes for cellulosic fibres are applied under neutral to alkaline conditions, since only at high pH values are cellulosic fibres charged. Direct dyes are large molecules which can have high affinity for the cellulosic fibres. The presence of sulphonic groups in their structure enhances solubility but necessitates the use of salts to balance negative charges on the fibre and on the dye molecules. Reactive dye molecules have a reactive head (vinylsulphonic or halotriazine groups) which reacts with cellulose at high pH, but like direct dyes, they have a similar need for salts. Vat dyes are insoluble and must be reduced to be soluble in water, at which stage they can be adsorbed onto the fibres with high salt concentrations and later reoxidised by air or hydrogen peroxide, as they are trapped inside the fibres.

Dyeing protein fibres is performed at neutral to acidic pHs. Acid dyes are easily adsorbed and fixed in wool fibres but with low fastness. Increased fastness can be achieved with metallic dyes that have large structures or with mordent dyes containing chromium which creates extra bonds between the dye and the fibre. Reactive dyeing of wool is performed under acid conditions using reactive groups similar to those in the reactive dyes for cotton.

Dyeing synthetic fibres is performed at temperatures above the glass transition, when dyes can penetrate better inside the fibre. The mechanism could be called 'solubilisation' of the dye inside the fibre, with high fastness resulting. Cationic dyes are linked to acrylics owing to the existence of negative groups on acrylics such as comonomers with sulphonic groups. Polyamides can be dyed with acid dyes because of the existence of amide groups that charge positively and can fix anionic dyes. If a white colour is demanded the fabrics can be delivered with double bleaching only, but if super whites are desired, optical brighteners can also be added to the fabric.

3.4.1.3 After coloration

Processes after coloration may include a variety of chemical and mechanical treatments where an effect can be added to or removed from the fabric: dimensional stability treatments, anti-crease finishing, softening, sanforisation, calendering, lamination, carding and others.

3.4.2 Desizing cotton with amylases

The use of α-amylases for desizing starch and their derivatives from woven fabrics was introduced almost 100 years ago. The enzymes used are mainly of bacterial origin such as *Bacillus subtilis*. Owing to advances in biotechnology a range of amylases acting at different temperatures from 20°C up to 115°C is available today. The optimum pH of the treatment lies between 5 and 7, depending on the enzymes used. All kinds of techniques can be used for the treatment ranging from padding to exhaustion methods. Amylases are used to desize fabrics made of dyed yarns, where oxidative desizing agents cannot be applied. Enzymatic desizing is the method of choice in wetting processing routes prior to dyeing when high levels of dye fastness are demanded, owing to the fast and very efficient removal of starch. Incomplete removal of starch might cause friction fastness problems.

3.4.3 Enzymatic removal of H_2O_2

Catalases were successfully introduced to the textile industry for the removal of hydrogen peroxide after bleaching and prior to dyeing at the beginning of the 1990s. The fast decomposition of hydrogen peroxide by catalases leads to a reduction in water consumption during washing the bleached cotton and prevents problems in further dyeing. For some catalases the pH of bleaching or washing liquors has to be adjusted to neutral values. Catalases are multimeric enzymes that might lose their activity in the presence of some surfactants by denaturating the fourth level structure of the enzyme, which should be considered in the choice of bleaching compositions (Costa *et al.*, 2001).

3.4.4 Cellulase finishing

Cellulases are the most successful enzymes used in textile processing. They can be used to obtain an aged or renewed look for cotton fabric. Cellulase systems include the individual enzymes endoglucanases (EGs) and cellobiohydrolases. For the generation of ageing effects EGs or EG-rich mixtures are used, while for renewal and depilling effects complete mixtures can be applied. Commercially available cellulases are mainly produced from the fungi *Humicola insolens* (optimum activity at pH 7) and *Trichoderma reesei* (optimum activity at pH 5). Although monocomponent EGs and EG-enriched products have been made available recently and have proved to be successful in many applications, for economic reasons mainly cellulase mixtures are still used (Cavaco-Paulo, 1998).

3.4.4.1 Depilling/cleaning effects

Fabric or garment depilling is usually carried out after heavy processing where pills are raised. Cellulases are used for pilling removal from fabric surfaces in machinery with high levels of mechanical agitation like jets, winches or drum washing machines. The most likely mechanism of enzymatic depilling/cleaning is the action of the enzyme (adsorption/hydrolysis) on easily accessible pills (or fibrils) at the surface of a fabric (or fibre). The pills become weaker after partial hydrolysis by cellulase and they are removed from the fabric by mechanical action. This mechanism is supported by the fact that depilling effects only take place at higher levels of mechanical agitation.

3.4.4.2 Ageing effects

The action of cellulases and mechanical agitation, simultaneously or sequentially, will abrade fibre surfaces, releasing cotton powder and causing defibrillation at the surface. In denim fabrics, because of enzymatic abrasion dye or dye aggregates with cotton will be released from yarns giving contrasts in the blue colour. The fibrillation produced during the ageing process is a result of the synergistic action of cellulases and mechanical action, and therefore the aged look is produced by less abrasive methods than traditional washing with pumice stones. This is the main advantage of the enzymatic washing process.

3.4.4.3 Key features of cellulase processing

In both applications mechanical agitation is very important as it seems to create more sites for cellulase attack either because of increased diffusion into the fabric or due to the increased surface area after defibrillation. Prior to direct and reactive dyeing, hard water and high ionic strength buffers negatively influence the performance of cellulases. Similarly, ionic surfactants inhibit cellulases. Dyeability and moisture recovery are not expected to change after cellulase treatment, since no changes occur in crystallinity of cellulase-treated cotton. However, owing to defibrillation, water retention has been shown to increase. Sometimes, slightly deeper shades are apparently obtained after cellulase treatment that cleans fibre surfaces.

Experimental evidence using complete crude mixtures and EG-enriched compositions suggest that strength loss is mainly produced by EG activity.

3.4.4.4 Indigo backstaining during enzymatic washing

The redeposition of the removed indigo dye by washing on the reverse side of denim is commonly known as backstaining. A mechanism responsible for backstaining has been proposed (Andreaus *et al.*, 2000) which suggests that cellulase proteins interact with indigo, reducing indigo particle size and acting as carriers of fine indigo particles already dispersed in the bulk solution to the cotton fabric. Since cellulases adsorb and desorb continuously during their hydrolytic activity on cotton cellulose (Azevedo *et al.*, 2000), it can be expected that cellulase proteins function as carriers of microfine indigo particles. After enzyme desorption from the cotton fabric indigo particles remain attached to the cellulosic fibres. In fact, cellulase enzymes can carry up to 250 times their weight in delivering other materials to cellulosic fabrics (Jones and Perry, 1998). The adsorption of indigo onto cellulases and the capacity for carrying microfine indigo particles depends on the type of the enzyme and the presence and type of the cellulose binding domain of the enzyme used. The best way to reduce backstaining is to perform a good wash after stone-washing, independent of the enzyme used.

3.4.4.5 Cellulosic fibres

Cellulosic fibres are currently the only 'synthetic' fibres treated with enzymes. Cellulase dosages applied to regenerated cellulose fibres are lower than for cotton as the former fibres are more susceptible to enzyme attack. This is mainly due to the fact that regenerated cellulose is present as cellulose II. In the area of synthetic fibres, cellulases are mainly used for the treatment of lyocell fabrics having a high pilling tendency after processes with strong mechanical agitation. Cellulases are essential finishing agents when used in a processing route to obtain a peach-skin feeling. When lyocell fabrics are subjected to a process with strong mechanical action, so-called primary fibrillation is produced (with raised longer fibres and fibrils). Cellulases can be used to clean fabric and fibre surfaces; thereafter another treatment with high mechanical action is applied and a secondary and uniform fibrillation is produced with very short fibrils, giving the peach-skin feeling.

3.4.5 Enzymes in detergents

Detergents are one of the most important markets for industrial enzymes (Ee *et al.*, 1997, Cavaco-Paulo, 1998). The function of the enzymes in detergents is to enhance the removal of soil particles by breaking them into smaller particles which can be more efficiently washed off.

Proteases have been used since the late 1960s in fabric washing products. Unspecific enzymes are used to work on a variety of protein soils. This implies that wool and silk fabrics cannot be washed with detergent formulations containing proteases. However, under mild washing conditions and short treatment times, little or no damage is produced in these fabrics. Lipases are also used in some detergent formulations to hydrolyse fats, improving detergency of fat soils. However, the benefit of these enzymes is still under discussion. Lipases seem to adsorb on fat soils and degradation occurs between the washing steps, giving complete removal in the subsequent wash. Amylases are also part of a few detergents that remove starch soils.

Cellulases were claimed to aid detergency during fabric washing more than 30 years ago. The known effect of microfibril removal by cellulases will help to liberate entrapped soils at disrupted fibre surfaces. The cleaning of fibre surfaces from soils and loss of microfibrils will give a brighter effect to fabrics and garments making garments look renewed. However, the first cellulases available were not active enough at alkaline pH values during washing. Nowadays, alkaline cellulase preparations containing mainly EGs are available and are used in detergents.

3.5 Promising areas of enzyme applications in textile processing

3.5.1 Enzymatic scouring of cotton

Scouring cotton with enzymes is one of the areas where considerable research effort has been expended resulting in the release of a commercial product. In these studies, lipases, pectinases, proteases, cellulases and their mixtures were used to improve cotton properties. Contradictory statements are reported in the literature about the efficiency of a new bioscouring formulation based on a pectin lyase. The major advantages feasible with this product seem to be savings in water and energy consumption, since the process is carried out at milder pH values and at lower temperatures when compared with traditional boiling scouring processes.

3.5.2 Bleaching

Bleaching processes for cotton have been proposed based on the application of glucose–oxidase for controlled production of hydrogen peroxide during oxidation of glucose released during enzymatic desizing. The resulting gluconic acid has been reported to serve as a sequestering agent for metal ions (Fe III) (Tzanov *et al.*, 2001).

Laccases have been suggested as a pretreatment step for subsequent peroxide bleaching to achieve high levels of whiteness. In the future, this process might replace two consecutive peroxide bleaching steps for bleaching cotton or flax fibres (Tzanov *et al.*, 2002a). Furthermore, a laccase mediator system has been launched on the market recently for bleaching denim fabrics. However bleaching levels are still low when compared with traditional agents like hydrogen peroxide.

3.5.3 New finishing enzymes for cotton

Permanent-press finishing with crosslinking agents generally induces fabric strength loss. With crosslinking agents such as polycarboxylic acids and *N*-hydroxymethyl acryl amide, ester and amide bonds are formed after curing. A controlled enzymatic hydrolysis with lipases and proteases has shown an increase in fabric strength without the loss of the permanent-press proprieties (Tzanov *et al.*, 2002b and Stamenova *et al.*, 2003). New antiflammatory properties have also been induced on cotton by a treatment with hexokinases and adenosine triphosphate (ATP), with phosphate groups being attached at the C6 of the glucosidic units at the surface of cellulose (Tzanov *et al.*, 2002c).

3.5.4 Lignocellulosic fibres

Bast fibers (flax, hemp, jute, kenaf and others) are composed of cellulose (over 50%), hemicelluloses, lignin, pectins, fats, waxes and others substances. Bast fibres are extracted from the plant stem by a process called 'retting' as mentioned before. The purpose of retting is the partial degradation of the fibre materials, in such a way that fibres can be obtained from the plant stems. Former retting processes of flax were based on incubation with bacteria and moisture (the stem in an open grass field) or in water (immersing the stem in slow rivers); nowadays retting is more often carried out in tanks of water at 30°C. Despite being an old process, much attention has been recently given to retting. The use of enzymes like hemicellulases and pectinases for retting allows a more controlled degradation of the fibres and a reduction of effluents. The up-grading of bast fibres is based on the use of cellulases for cleaning and softening, simplifying further processing. However, in retting or further softening treatments, care should be taken since the removal of fibrous material may yield unacceptable levels of strength loss (Cavaco-Paulo, 1998).

3.5.5 Wool processing with enzymes

Most of the wet processing steps of wool are carried out under very mild agitation owing to the tendency of wool to felt. This tendency is a result of the presence of the 'scales' of the cuticles on the wool surface. The removal or modification of these scales by oxidation and addition of polymers causes antishrinking behaviour. Most of the chemicals used for oxidation (halogen derivatives) are environmentally harmful and therefore, intensive research has been made to develop more environmentally friendly processes. The investigation of enzymatic processes for antishrink finishing of wool dates back to 1910, when trypsin and pepsin were used to clean skin scales. The first studies showed that the preswelling of the fibre could determine the extent of proteolysis. It was also stated that, if the cystine disulphide bonds remained intact, the proteolysis was slow. However, when some of the crosslinks were broken, the reaction rate increased. Several processes (already patented in the 1940s) based on an oxidative treatment followed by proteolysis had been suggested, but none was applied to the industry because of the high enzyme costs and the unacceptable weight losses obtained. Reports about the use of papain and commercial proteases after oxidative treatment, showed a good 'descaling' effect but high fibre damage. Various studies suggest that enzymes affect mainly the inner part of wool, confirming that the enzymes seem to diffuse inside the fibre, 'retting' it (Cavaco-Paulo, 1998).

The use of protein disulphide isomerase has been reported to improve the shrinkage behaviour of wool fabrics. This enzyme rearranges disulphide bonds with the aid of a cofactor in a reduced form, such as glutathione or dithiothreitol. The use of transglutaminase has also been reported to improve shrinkproofing of wool, by a rather different mechanism, with the formation of new crosslinks (*N*6-(5-glutamyl)-lysine)) and the liberation of ammonia. Attempts to replace carbonisation of wool by enzyme treatments have been made using a range of different enzymes to remove vegetable matter, reducing the amount of sulphuric acid used. However, the enzymatic degradation of vegetable materials by hydrolases such as cellulases and hemicellulases is a slow process (Cavaco-Paulo, 1998).

3.5.6 Enzymes on silk

Silk fibres are composed mainly of a double filament of fibroin surrounded by a layer called sericin. Both sericin and fibroin are proteins that have almost no cysteine residues after hydrolysis. The amount of sericin, in terms of weight loss after degumming varies between 17–38%. Sericin is mainly composed of serine (33%), aspartic acid (17%), glycine (14%) and minor quantities of other residues. Fibroin is mainly composed of glycine (44%) and alanine (29%). Sericin is more accessible to chemicals than fibroin and it is removed during preparation. An ideal degumming agent would specifically attack peptide bonds near serine residues. However, several methods have been developed for degumming silk, such as extraction with water, boiling with detergent, with alkali, acids and enzymes. Degumming with commercially available bacterial proteases is more effective than using trypsin and papain. Proteases can also be used to alter the silk fibroin surface to give an aged look in a similar way to enzyme washing of denim garments. It has also been recently reported that proteases may provide improved softness and wetability of silk fabrics.

3.5.7 Synthetic fibres

Several enzymes can catalyse the modification of synthetic polymers. The modification of polyacrylonitrile with enzymes (nitrile hydratase) increases the number of amide groups on the fibre surface giving improved dyeability and hydrophilicity. The same enzymes are also used for production of precursors of polyamide 6,6. This could be the beginning of the application of enzymes in the production of synthetic textile fibres. Esterases and peptidases are good candidates for treatment of polyester and polyamide, but no processes have been introduced so far in the textile industry.

Acknowledgements

We thank Dr. Herman Lenting for his critical reading of the chapter and Barbara Klug for her insight into the section about pectinases.

3.6 References

Abadulla E., Tzanov T., Costa S., Robra K., Cavaco-Paulo A. and Gübitz G. (2000) 'Decolorization and detoxification of textile dyes with a laccase from *Trametes hirsuta*', *Appl. Environ. Microbiol.*, **66**, 3357–3362.

Andreaus J., Campos R., Gübitz G. and Cavaco-Paulo A. (2000) 'Influence of cellulases on indigo back staining', *Textile Res. J.*, **70**, 628.

Atkins P. and Paula J. (2001) *Physical Chemistry*, Oxford University Press, 7th Edition, UK.

Azevedo H., Bishop D. and Cavaco-Paulo A. (2000) 'Effects of agitation level on the adsorption, desorption and activities on cotton fabrics of full length and core domains of EGV (*Humicola insulens*) and CenA (*Cellulomonas fimi*)', *Enzyme Microbial Technol.*, **27** (3–5), 325–329.

Bailey C.J. (1989) 'Enzyme kinetics of cellulose hydrolysis', *Biochem. J.*, **262**, 1001.

Bergmeyer H.U. (ed) (1974) *Methods of Enzymatic Analysis*, 3rd Edition, Academic Press, New York.

Beynon R.J. and Bond J.S. (1996) *Proteolytic Enzymes: a practical approach*, Oxford University Press, Oxford.

Biely P., Benen J., Heinrichová K., Kester H.C.M. and Visser J. (1996) 'Inversion of configuration during hydrolysis of α-1,4-galacturonidic linkage by three Aspergillus polygalacturonases', *FEBS Lett.*, **382** (3), 249–255.

Bisswanger H. (2002) *Enzyme Kinetics: principles and methods*, Wiley-VCH, Weinheim.

Cambillau C., Longhi S., Nicolas A. and Martinez C. (1996) 'Acyl glycerol hydrolases: inhibitors, interface and catalysis', *Curr. Opinions Struct. Biol.*, **3**, 449–455.
Cavaco-Paulo A. (1998) *Processing Textile Fibres with Enzymes: An overview*, ACS Symposium Series, **687**, 180–189.
Cavaco-Paulo A., Almeida L. and Bishop D. (1996) 'Cellulase activities and finishing effects', *Textile Chem. Color.*, **28** (6), 28–32.
Cavaco-Paulo A., Almeida L. and Bishop D. (1998) 'Hydrolysis of cotton cellulose by engineered cellulases from *Trichoderma reesei*', *Textile Res. J.*, April, 273–280.
Cavaco-Paulo A., Morgado J., Andreaus J. and Kilburn D. (1999) 'Interactions of cotton with CBD peptides', *Enzyme Microbial Technol.*, **25** (8–9), 639–643.
Costa S., Tzanov T., Paar A., Gudelj M., Gübitz G. and Cavaco-Paulo A. (2001) 'Immobilization of catalases from *Bacillus* SF on alumina for the treatment of textile bleaching effluents', *Enzyme Microbial Technol.*, **28**, 815–819.
Danson M. and Hough D. (1998) 'Structure, function and stability of enzymes from the *Archaea*', *Trends Microbiol.*, Aug;**6** (8), 307–314.
Davies G. and Henrissat B. (1995) 'Structures and mechanisms of glycosyl hydrolases', *Structure*, **3**, 853–859.
Ee J., Misset O. and Baas E. (1997) *Enzymes in Detergency*, Surfactant Science Series, 69, Mercel Dekker, New York.
Erkkilä M., Leah R., Ahokas H. and Cameron-Mills V. (1998) 'Allele-dependent barley grain beta-amylase activity', *Plant Physiol.*, **117** (2), 679–685.
Ferreyra O.A., Cavalitto S.F., Hours R.A. and Ertola R.J. (2002) 'Influence of trace elements on enzyme production: protopectinase expression by a *Geotrichum klebahnii* strain', *Enzyme Microb. Technol.*, in press.
Ghose T.K. (1987) 'Cellulase activities', *Pure Appl. Chem.*, **59**, 257–268.
Gianfreda L., Xu F. and Bollag J.-M. (1999) 'Laccases: A useful group of oxidoreductive enzymes', *Bioremediat. J.*, **3** (1), 1–26.
Gudelj M., Fruhwirth G.O., Paar A., Lottspeich F., Robra K.H., Cavaco-Paulo A. and Gübitz G.M. (2001) 'A catalase-peroxidase from a newly isolated thermoalkaliphilic Bacillus sp. with potential for the treatment of textile bleaching effluents', *Extremophiles*, **5** (6), 423–429.
Hakulinen N., Tenkanen M. and Rouvinen J. (2000) 'Three-dimensional structure of the catalytic core of acetylxylan esterase from *Trichoderma reesei*: Insights into the deacetylation mechanism', *J. Struct. Biol.*, **132**, 180–190.
Henrissat B. (1991) 'A classification of glycosyl hydrolases based on amino-acid sequence similarities', *Biochem. J.*, **280**, 309–316.
Henrissat B. and Bairoch A. (1993) 'New families in the classification of glycosyl hydrolases based on amino-acid sequence similarities', *Biochem. J.*, **293**, 781–788.
Henrissat B. and Bairoch A. (1996) 'Updating the sequence-based classification of glycosyl hydrolases', *Biochem. J.*, **316**, 695–696.
Herron S., Benen J., Scavetta R., Visser J. and Jurnak F. (2000) 'Structure and function of pectic enzymes: Virulence factors of plant pathogens', *Proc. Nat. Acad. Sci. USA*, **97** (16), 8762–8769.
Horvathova V., Janecek S. and Sturdik E. (2001) 'Amylolytic enzymes: molecular aspects of their properties', *Gen. Physiol. Biophys.*, **20** (1), 7–32.
Jones C.C. and Perry A. (1998) *Detergent composition*, PCT Intl. Appl., WO 98/00500, 38 pp.

Kobayashi M. and Shimizu S. (1998) 'Metalloenzyme nitrile hydratase – structure, regulation, and application to biotechnology', *Nat. Biotechnol.*, **16**, 733–736.

Lee S., He S. and Withers S. (2001) 'Identification of the catalytic nucleophile of the Family 31 α-glucosidase from *Aspergillus niger* via trapping of a 5-fluoroglycosyl–enzyme intermediate', *Biochem. J.*, **359**, 381–386.

Machius M., Declerck N., Huber R. and Wiegand G. (1998) 'Activation of *Bacillus licheniformis* alpha-amylase through a disorder–order transition of the substrate-binding site mediated by a calcium–sodium–calcium metal triad', *Structure*, **6** (3), 281–292.

Matsumoto T., Sugiura Y., Kondo A. and Fukuda H. (2000) 'Efficient production of protopectinases by *Bacillus subtilis* using medium based on soybean flour', *Biochem. Eng. J.*, **6**, 81–86.

Numao S., Maurus R., Sidhu G., Wang Y., Overall C.M., Brayer G.D. and Withers S.G. (2002) 'Probing the role of the chloride ion in the mechanism of human pancreatic R-amylase', *Biochemistry*, **41**, 215–225.

Pace H. and Brenner C. (2001) 'The nitrilase superfamily: classification, structure and function', *Genome Biol.*, **2** (1), reviews 0001.1–0001.9.

Palmer T. (1995) *Understanding Enzymes*, Ellis Horwood, 4th Edition, UK.

Petersen M., Fojan P. and Petersen S. (2001) 'How lipases and esterases work: electrostatic contribution', *J. Biotechnol.*, **85**, 115–147.

Sakai T., Sakamoto T., Hallaert J. and Vandamme E. (1993) 'Pectin, pectinase and protopectinase: production, properties and applications', *Adv. Appl. Microbiol.*, **39**, 213–294.

Sakamoto T. and Sakai T. (1994) 'Protopectinase-T: a rhamnogalacturonase able to solubilize protopectin from sugar beet', *Carbohydrate Res.*, **259** (1), 77–91.

Schulein M. (1997) 'Enzymatic properties of cellulases from *Humicola insolens*', *J. Biotechnol.*, **57**, 71–81.

Stamenova M., Tzanov T., Betcheva R. and Cavaco-Paulo A. (2003) 'Proteases to improve the mechanical characteristics of durable press finished cotton fabrics', *Macromol. Materials Eng.*, **288**, 71–75.

Tauber M., Cavaco-Paulo A., Robra A. and Gübitz G. (2000) 'Nitrile hydratase and amidase from *Rhodococcus rhodochrous* hydrolyse acrylic fibers and granulates', *Appl. Environ. Microbiol.*, **66**, 1634–1638.

Tomme P., Kwan E., Gilkes N.R., Kilburn D.G. and Warren R.A. (1996) 'Characterisation of CenC, an enzyme from *Cellulomonas fimi* with both endo- and exoglucanase activities', *J. Bacteriol.*, **178**, 4216–4223.

Tomme P., Warren A.J. and Gilkes N.R. (1995) 'Cellulose hydrolysis by bacteria and fungi', *Adv. Microbial Physiol.*, **37**, 1–87.

Tzanov T., Costa S., Gübitz G. and Cavaco-Paulo A. (2001) 'Immobilized glucose oxidase for hydrogen peroxide generation for textile bleaching', *J. Biotechnol.*, **93** (1), 87–94.

Tzanov T., Gübitz G. and Cavaco-Paulo A. (2002a) *Pré-tatramento com lacares para-methorer o grau de branco de materials tèteis*, Portuguese Patent Application No. 102779, in 2002.03.15.

Tzanov T., Stamenova M., Betcheva R. and Cavaco-Paulo A. (2002b) 'Lipases to improve the performance of formaldehyde-free durable pressfinished cotton fabrics', *Macromol. Materials Eng.*, **278**, 462–465.

Tzanov T., Stamenova M. and Cavaco-Paulo A. (2002c) 'Phosphorylation of cotton cellulose with baker yeast hexokinase', *Macromol. Rapid Commun.*, **23**, 962–964.

Watanabe K., Miyake K. and Suzuki Y. (2001) 'Identification of catalytic and substrate-binding site residues in *Bacillus cereus* ATCC7064 oligo-1,6-glucosidase', *Biosci. Biotechnol. Biochem.*, **65** (9), 2058–2064.

Whitaker J.R. (1989) 'Microbial pectolytic enzymes', in *Microbial Enzymes and Biotechnology*, 2nd Edition, eds Fogerty W. and Kelly C., Elsevier, London, pp. 133–175.

Wood, M.T. (1992) 'Fungal cellulases', *Biochem. Soc. Trans.*, **20**, 46–53.

Woodward J., Lima M. and Lee N.E. (1988) 'The role of cellulase concentration in determining the degree of synergism in the hydrolysis of microcrystalline cellulose', *Biochem. J.*, **255**, 895–899.

Zamocky M., Regelsberger G., Jakopitsch C. and Obinger C. (2001) 'The molecular peculiarities of catalase-peroxidases', *FEBS Lett.*, **492** (3), 177–182.

4
Process engineering and industrial enzyme applications

V. A. NIERSTRASZ AND
M. M. C. G. WARMOESKERKEN
University of Twente, The Netherlands

4.1 Introduction

Biocatalysis plays an increasingly important role in industrial wet textile pretreatment and finishing processes. Conventional wet textile processes are characterised by long residence times, high concentrations of chemicals, alkaline or acidic pH and high temperatures. It is to be expected that wet textile processes will be shifted considerably towards sustainable processes based on biocatalysis, owing to increasing governmental and environmental restrictions and the decreasing availability of fresh water. Biocatalysis is a flexible and reliable tool that presents a promising technology for fulfilling expected future requirements.

Since the early 1990s a lot of research has been done on reactions catalysed by enzymes that are relevant to the textile industry. Often these studies focus on the enzymatic incubation itself and the enzyme and substrate characteristics and not on parameters necessary for the design of efficient and competitive full-scale industrial processes.

Process parameters need to be related to cloth properties such as the porosity and the density of the fabric in order to introduce efficient and economic enzymatic treatments. The design of enzymatic textile treatment processes is a difficult task, which is often based on trial and error instead of process engineering. On the one hand this is caused by the complex geometrical structure of textile materials and on the other hand by the specific kinetics of enzymatic reactions and the relatively large sizes of enzyme molecules. The time-determining step in the kinetics of these processes is often the transport of molecules to the surfaces of the textile fibres. Although the thickness of textiles is small, in many cases less than 1 mm, the porous structure of the material hinders a free flow of liquid. This means that diffusion of molecules through the pores to the fibre surfaces is the main transport mechanism. This is a relatively slow process, especially if the diffusing molecules are large like enzymes. Therefore much time is needed before the enzyme molecules are adsorbed at all fibre surfaces. It also takes a considerable time before all the reaction products have been removed from porous textiles. In order to say something about the residence time of the fabric in an impregnation step or in a washing out step, the diffusion time of the enzymes in textiles has to be determined.

This chapter gives an overview of different industrial enzymatic cotton pretreatment and finishing processes and focuses on mass transfer in enzymatic wet textile processes. Different possibilities for process intensification are considered and, as well as mass transfer limitation in immobilised enzyme systems, an application of biocatalysis especially relevant in the treatment of the effluents of textile mills is discussed.

4.2 Large-scale industrial enzyme applications in textiles: an overview

The estimated value of the world enzyme market was about US $1.5 billion in 2000 and it has been forecasted to grow to US $2 billion in 2005. In Tables 4.1 and 4.2 large-scale industrial enzyme applications and their market size are summarised (Rehm *et al.*, 1996).

Detergents, textiles, food, starch, paper and pulp, baking and animal feed are the main industries that use approximately 75% of the industrially produced enzymes. The largest manufacturers of industrial enzymes are Novozymes, Genencor, DSM and Röhm & Haas. Detergents have always been the largest application of industrial enzymes. Inventions made in the field of enzyme applications in detergents quite often found their application later in the textile industry. Detergents were also the first large scale application for microbial enzymes. Röhm in Germany had already produced the first commercial enzyme used in a detergent in 1914. Bacterial proteinases are the most important detergent enzymes. Some products have been produced by genetically modified organisms to be more stable or

Table 4.1 Market size of large-scale industrial enzyme applications

Industry	Market size (10^6 US $)
Detergent	500
Textile	150
Drinks/brewing	150
Dairy	150
Pulp and paper	100
Starch	100
Baking	100
Animal feed	80

Table 4.2 Large-scale industrial enzyme applications

Enzyme	Application	Industry	Market size (%)
Protease	Protein degradation	Detergent	50
α-Amylase	Glucose production	Starch	16
	Desizing	Textile	
Cellulase	Colour brightening	Detergent	14
	Fibril removal	Textile	
	Juice extraction	Drinks	
α-Amylase	Shelf life	Baking	11
Lipase	Fat removal	Detergent	7
Pectinase	Juice clarification	Drinks	4
	Scouring	Textile	

active in the hostile environment of the washing machine or the detergent, high temperatures, alkaline pH and oxidising agents.

In the late 1980s lipid-degrading enzymes were introduced in powder and liquid detergents. Lipases hydrolyse ester bonds of fats, thereby producing glycerol and fatty acids. Amylases are used in detergents to remove starch-based stains that stick on textile fibres and bind other stains. Cellulases were introduced in detergents in the early 1990s. Cellulases are able to degrade cellulose and are therefore able to remove cellulose microfibres that are formed during the use and washing of cotton products. The removal of these microfibres by cellulases results in colour brightening and softening of the textile material. Recent developments are in the field of thermostable enzymes, protein engineering and enzymes obtained by genetically modified microorganisms.

The use of enzymes in the textile industry is one of the most rapidly growing fields in industrial biocatalysis (Thiry, 2001). For a long time starch has been used as a protective lubricant and glue of fibres in the weaving of fabrics; amylases are used to remove the starch in the desizing process. Cellulases are used for the enzymatic depilling of cotton fabrics and in the production of denim fabrics. The fading effect on indigo-dyed cotton used to be created by pumice stones, but the pumice stones caused damage to both fibres and machines. The same fading effect is nowadays obtained with cellulase enzymes. An application introduced more recently in the textile industry is the use of enzymes in the cotton scouring process. During scouring, waxes and other hydrophobic material are removed from the cotton fibres. Conventionally this process is done in hot sodium hydroxide (NaOH). Alkaline pectinases are able to degrade pectin in the outer layers of the fibre, thereby weakening the structure of the outer layers so they can be removed afterwards.

4.3 Industrial applications of enzymes in wet textile processing

Enzymes are gaining an increasingly important role as a tool in various wet textile pretreatment and finishing processes (Stanescu, 2002; Thiry, 2001; Cavaco-Paulo, *et al.* 1998; Heine and Höcker, 1995). Conventional wet textile pretreatment and finishing procedures applied in the textile industry are often characterised by high concentrations of chemicals, alkaline or acidic pH, and high temperatures with consequent high consumption of energy. Enzymes are very specific catalysts; they operate best at ambient pressures, mild temperatures and often at a neutral pH. It is to be expected that, within 5 to 10 years, wet textile production processing will be shifted substantially towards sustainable processes, because of increasing governmental and environmental restrictions and the decreasing availability of fresh water. Biocatalysis has proven to be a flexible and reliable tool in wet textile processing and a promising technology for fulfilling expected future requirements.

In the scientific literature a lot of detailed information can be found on the different reactions catalysed by enzymes that are relevant to the textile industry, such as desizing, biopolishing, biostoning and more recently bioscouring (for more examples see preceding chapters). Most studies described in scientific literature focus on aspects that are directly related to enzymatic incubation, the enzyme and substrate characteristics (e.g. Agrawal *et al.*, 2002; Buchert *et al.*, 2000; Buschle-Diller *et al.*, 1998; Cavaco-Paulo *et al.*, 1996, 1997, 1998b; Etters, 1999; Hartzell and Hsieh, 1998; Lenting and Warmoeskerken, 2001a, 2001b; Lenting *et al.* 2002; Li and Hardin, 1998; Pere *et al.*, 2001; Tzanov *et al.*, 2001; Yachmenev *et al.*, 2001). However, apart from reaction mechanisms, the relationship between substrate and enzyme, the amount of shear or agitation, optimal temperature and pH etc., these studies often do not focus on parameters necessary for the design of true full-scale industrial processes. This is partially caused by the fact that mass transfer and shear, for example, are quite different in laboratory-scale equipment than in industrial batch and (semi-)continuous equipment. Process parameters need to be related to cloth properties such as the porosity and the density of the fabric in order to introduce efficient and economic enzymatic treatments. Most information relevant for the design and development of industrial processes comes from companies producing enzymes or companies that develop formulations and applications for the textile industry (e.g. Bayer, Genencor, Novozymes, Dexter Chemical Corp.) and some from scientific or more technical publications that are dedicated to industrial enzymatic wet textile pretreatment or finishing processes (Lange and Henderson, 2000; Lange, 2000; Cortez *et al.*, 2001, 2002; Contreras, 2001; Waddell, 2002).

4.3.1 Desizing of cotton

During weaving, warp yarns are exposed to considerable mechanical strains. To prevent the yarns from breaking, they are coated with a sizing agent (a protective glue and lubricant). The sizing agent is most often based on starch. Apart from starch, synthetic sizing agents are available, for example polyvinyl alcohol (PVALc), but, for economic reasons, starch is still the most favourable sizing agent. After weaving the fabric, the sizing agent needs to be removed since it hinders textile-finishing processes such as dyeing. This desizing process used to be done chemically using, for example, hydrogen peroxide, (H_2O_2) and sodium hydroxide (NaOH), but since the 1950s enzymatic desizing processes based on α-amylases have been widely introduced and implemented successfully in the textile industry. In the enzymatic desizing process an almost complete removal of starch-containing size is obtained without any fibre damage. Amylases were derived from mulds or pancreas but are nowadays produced by bacteria (especially *Bacillus subtilis*). Biotechnological progress and genetic engineering of microorganisms have allowed thermostable enzymes to be widely available nowadays, and therefore different temperatures, ranging from 20 to 80°C for conventional α-amylases and 40 to 110°C for thermostable α-amylases, are applicable. The optimum pH lies between 4 and 10 depending on the enzyme used. Most α-amylases are suitable for all common batch and (semi-)continuous processes, such as jet, jig, cold and hot pad-batch (see Fig. 4.1), pad-steam and J-box. Typical process conditions for woven cotton fabrics are summarised in Table 4.3 (data from product guides and product information sheets from Novozymes, Bayer and Genencor).

During impregnation the hot water causes the starch to gelatinise and the fabric becomes fully wetted and impregnated with the enzyme solution. The average liquid take-up during impregnation is approximately 1 L/kg

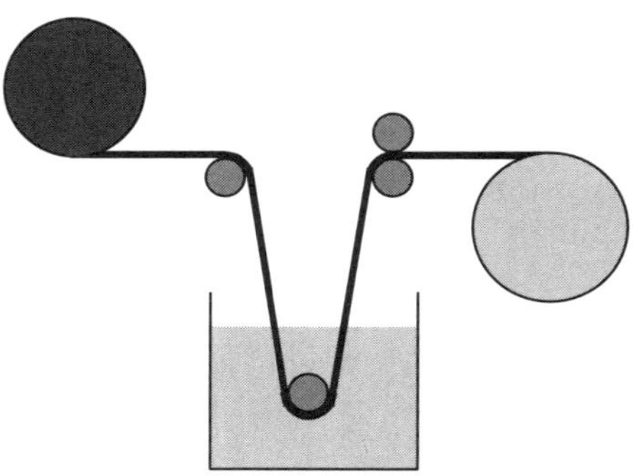

4.1 Pad-batch process, typically used in the enzymatic desizing of cotton fabrics.

Table 4.3 Typical process conditions for some common textile desizing applications

	Jig	Winch	Pad-batch (cold)	Pad-batch (hot)	Pad-steam
Impregnation:					
Enzyme dosage (mL/L)	0.3–1	0.3–1	1–10	1–6	1–10
Temperature (°C)	60–95	70–100	15–40	60–70	20–110
pH	5–7.5	5–7.5	5–7.5	5–7.5	5–7.5
Incubation:					
Incubation time	2–4 (passages)	30 (min)	6–24 (h)	3–8 (h)	15–120 (s)
Temperature (°C)	60–100	90–100	15–40	60–70	90–110

fabric, and depends on the characteristics of the fabric such as the porosity and the additives present in the impregnation liquor. Chelating agents should preferably not be used during the desizing process because calcium ions (at ppm level) stabilise the enzymes. Wetting agents and non-ionic surfactants can be used to enhance enzyme penetration and adsorption, fibre swelling and to promote the removal of waxes, soils and synthetic sizing agents. Non-ionic surfactants are suitable for combination with enzymes, whereas anionic and cationic surfactant may inactivate the enzyme through denaturation. Lubricants are generally recommended to be used in combination with α-amylases during desizing, especially in jets and rotary washers, to reduce the formation of crease marks and streaks. After the enzymatic treatment, fabrics should be washed off above 80°C, often between 90 to 100°C, in alkaline liquor followed by a wash in neutral liquor.

4.3.2 Cotton finishing: enzymatic ageing and depilling

The application of cellulases in wet textile processes has been, like enzymatic desizing, successfully introduced and accepted in the textile industry. Cellulase enzymes are a class of hydrolytic enzymes that are used for different cotton finishing processes: cellulases can be utilised to give indigo-dyed cotton fabrics (denim) an aged appearance (also known as biostoning), and to give cotton fabrics a renewed appearance by colour brightening and softening of the material through the removal of microfibres (depilling, also known as biopolishing). As mentioned in Chapter 3, cellulase is a typical multicomponent enzyme that consists of mainly:

- Endoglucanases (EGs), (EC 3.2.1.4);
- Cellobiohydrolases (CBHs) or exo-cellobiohydrolases, (EC 3.2.1.91);
- β-glucosidases or cellobiases (EC 3.2.1.21).

Cellulases can be derived from a variety of microorganisms, especially fungi, such as *Trichoderma reesei*, *Humicola insolens*, *Aspergilus niger* and *Bacillus subtilus*. These organisms can all be used to produce acid-stable as well as neutral- and alkaline-stable cellulase mixtures. Natural cellulase mixtures are produced by microorganisms to hydrolyse insoluble cellulose very efficiently. In textile finishing processes this is not necessary or even undesirable.

Nowadays it is an accepted concept that the performance characteristics in textile finishing applications of a certain cellulase composition are determined by its specific composition, rather than the optimum pH or temperature of the enzymes present in the mixture, or the microorganism used to produce the enzymes. Besides conventional cellulase mixtures, dedicated cellulase compositions are nowadays available commercially, such as EG-enriched cellulase mixtures, monocomponent cellulases and even modified cellulase enzymes with unique performance features, thanks to modern biotechnological techniques.

4.3.2.1 Enzymatic ageing

The finishing of denim garments by pumice stones (stonewashed garments) to achieve an aged or worn appearance has been radically improved by the application of cellulase enzymes (also called cellulase washing or biostoning). This stonewash effect is due to abrasion of the fabric thereby locally removing the surface-bound indigo dye and revealing the white interior of the yarn. In the traditional stonewash process, abrasion is caused by pumice stones and by garments chafing against the washer drum. The pumice stones damage the washer drum and reduce the fabric strength due to abrasion. The application of cellulases prevents damage to the washing machine, eliminates the need for disposal of used stones and results in an improved quality of wastewater because of the absence of pumice stone dust. Owing to the absence of stones (several kilograms) the garment load may be increased up to 50% resulting in increased productivity. The use of cellulases results in a softer fabric and, because of the reduced abrasion of the fabric, an increased strength compared to the traditional stonewash process. The application of cellulases is, owing to the global market size for stonewashed denim garments, a very successful application of enzymes in the textile industry.

EG-enriched or EG monocomponent formulations are usually preferred because of their superior performance in biostoning (see the preceding chapter). Most commercially available formulations are produced from the fungi *Trichoderma reesei* (optimum pH 5) and *Humicola insolens* (optimum pH 6.5–7.0). EGs from *Trichoderma reesei* are known for their great effectiveness, and therefore give a flatter and lower contrast pattern and a small

Table 4.4 Typical process conditions for an industrial biostoning process

Parameter	
pH	4.5–7.0
Temperature	45–65°C
Liquid ratio	3:1–20:1
Incubation time	15–60 min

amount of hydrolysis of the fabric, but have a tendency to promote indigo backstaining (redeposition of indigo dye on the undyed white weft yarns of the fabric). Neutral EGs produced from *Humicola insolens* are known for their low levels of backstaining and the enzymes have a broader pH range. The latter property allows for a more reproducible process. Backstaining is promoted by adsorption of indigo on the enzyme and the subsequent adsorption of the enzyme on to the fabric. The use of enzymes with a low affinity for indigo and the absence of, for example, a cellulose binding domain (CBD) will thus result in a reduced amount of backstaining (see the preceding chapter). Different commercial formulations are nowadays available with specific key features for different applications and results.

An enzymatic stonewash process requires equipment with sufficient shear forces and mixing, such as a drum washer. Typical parameters for commercial formulations are summarised in Table 4.4 (Product Guide, Genencor). The use of a buffer solution is recommended, especially when applying an acidic formulation. Before enzymatic stonewashing, proper and complete desizing is recommended. The incubation time depends on the type of machine, the liquid ratio, the garments or fabrics and the desired effect or look. It is important not to overload the machine, because that will reduce the amount of shear force and mixing.

The enzyme dosage depends on the type, density, porosity and hydrophilicity of the fabric or garment and the effect desired. It is a function of treatment time, pH, temperature, liquid ratio, auxiliary chemicals and the type of equipment (shear force and mixing). In general, the addition of non-ionic surfactants and dispersing agents is recommended and will enhance the overall performance. The cellulase enzymes need to be inactivated after the desired stonewash effect is obtained. Insufficient inactivation will result in extended degradation of cellulose and therefore an undesirable strength loss and weight reduction. There are several options to inactivate the cellulases:

- Increase the pH (pH > 9) and raise the temperature ($T > 60$°C) for 15 minutes;

- Wash off the fabrics in an alkaline detergent solution (pH > 9 and $T > 60°C$) for 15 minutes;
- Perform a standard chlorine bleach of the fabrics.

4.3.2.2 Depilling

In cotton fabric, fuzz (microfibres) emerges from the surface. When these microfibres become entangled during processing, pills are formed. Depilling (also called biopolishing) is a cellulase treatment to improve the fabric quality, often done after heavy processing where pills are raised. In the enzymatic depilling of cellulosic fabrics such as cotton and Lyocell, these pills and fuzz are enzymatically removed (Cavaco-Paulo *et al.*, 1998). As in enzymatic ageing, EG or EG-enriched mixtures are most effective. Cellulase enzymes will weaken the fibres protruding from the surface by degradation, preferably of the amorphous structure of the fibre. The enzyme-weakened fibres are sensitive to shear forces and upon application of sufficient shear the fibre will break from the surface (Cavaco-Paulo *et al.*, 1996, 1997; Lenting and Warmoeskerken, 2001a). This results in:

- improved pilling resistance;
- brighter colours;
- cleaner surface;
- improved drapeability and increased softness;
- reduction in the amount of dead and immature cotton.

Enzymatic depilling is preferably carried out after bleaching the fabric, but can be carried out after any wet textile pretreatment step, after proper and complete desizing. Enzyme treatment after dyeing can result in partial dye removal and thus colour change depending on the dye used. An enzymatic depilling process requires equipment with sufficient shear forces and mixing such as a jet (see Fig. 4.2) or a winch (Cortez *et al.*, 2001).

Today's commercially available continuous equipment does not produce enough shear forces and mixing for enzymatic depilling. Typical process conditions for an industrial depilling process are pH 4.5–6.0, temperature 45–65°C, liquid ratio 3:1–20:1 and an incubation time of 15–60 minutes. The incubation time depends on the type of machine, the liquid ratio, the fabric and the desired effect. As in the denim ageing process, the enzyme dosage depends on the type, density, porosity and hydrophilicity of the fabric and the desired effect, and is a function of treatment time, pH, temperature, liquid ratio, auxiliary chemicals and the type of equipment (shear force and mixing). In general, non-ionic surfactants and dispersing agents are recommended and will enhance the overall performance. As in denim ageing, the cellulase enzymes need to be inactivated after the desired effect is obtained. Insufficient inactivation will result in extended degradation of

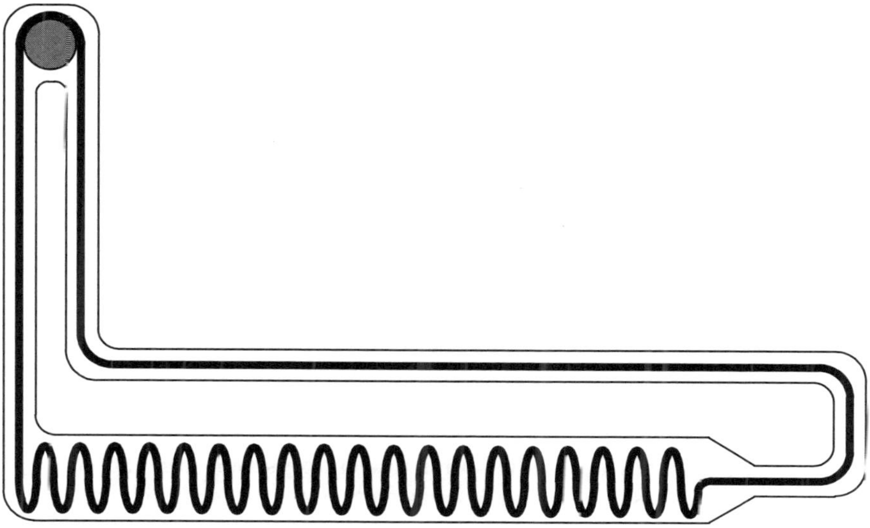

4.2 Jet, typically used for biopolishing processes.

cellulose and therefore an undesirable strength loss and weight reduction. Different suitable inactivation procedures can be found in the section on enzymatic ageing previously.

4.3.3 Scouring of cotton

Before grey cotton fabric can be dyed and finished it has to be treated in order to make it hydrophilic and to remove the primary cell wall (see preceding chapters). In conventional cotton scouring processes high temperatures (90–100°C) and high concentrations of NaOH (approx. 1 mol/L) are used to remove the primary cell wall (pectin, protein, organic acids) and hydrophobic components from the cuticle (waxes and fats) in a non-specific way to make the fibre hydrophilic. Owing to the high NaOH concentration, extensive washing and rinsing is required, causing increased water consumption. The use of high concentrations of NaOH also requires the neutralisation of the wastewater, which requires additional chemicals. It is obvious that this process needs to be improved considerably to meet today's energy and environmental demands. Much research has been directed to replace this process with an enzymatic one (see for example: Agrawal *et al.*, 2002; Buchert *et al.*, 2000; Buschle-Diller *et al.*, 1998; Csiszár *et al.*, 2001; Etters, 1999; Hartzell and Hsieh, 1998; Lenting *et al.*, 2002; Li and Hardin, 1998; Tzanov *et al.*, 2001; Yachmenev *et al.*, 2001). The potential to degrade and remove the undesired components from the cotton fibres of different enzymes, such as pectinases, cellulases and lipases, as well as different process conditions, have been investigated.

Novozymes, Bayer and Dexter Chemical Corporation have introduced an enzymatic alternative for scouring woven and knitted cotton fabrics in the textile industry on the basis of an alkaline pectinase (EC 4.2.2.2) produced by a genetically modified *Bacillus* strain. On an industrial scale, the bioscouring process using alkaline pectinases has been performed successfully in batch (pad-batch) and continuous (open width) processes (Lange and Henderson, 2000; Lange, 2000), and integrated with desizing in batch (pad-batch) and continuous (J-box) processes (Waddell, 2002). The idea is that pectin acts as a sort of cement or matrix that stabilises the primary cell wall of the cotton fibres. During incubation the enzymes will degrade pectin, thereby destabilising the structure in the outer layers. The weakened outer layers can be removed in a subsequent wash process. The bioscouring process results in textiles being softer than those scoured in the conventional NaOH process, however the degree of whiteness is often less and the process is not suitable for removing seed coat fragments and motes adequately.

A typical time–temperature profile for the enzymatic scouring of woven and knitted cotton fabrics in a jet machine is shown in Fig. 4.3 (data from Novozymes, Bayer, Lange and Henderson, 2000, Lange 2000 and Waddell, 2002). These conditions are applicable to most exhaustion machinery. The liquor ratio is 8:1 and the enzyme dosage should be between 0.5 and 1.0%.

Wetting agents and non-ionic surfactants should be added together with the enzyme to enhance enzyme penetration and adsorption, fibre swelling

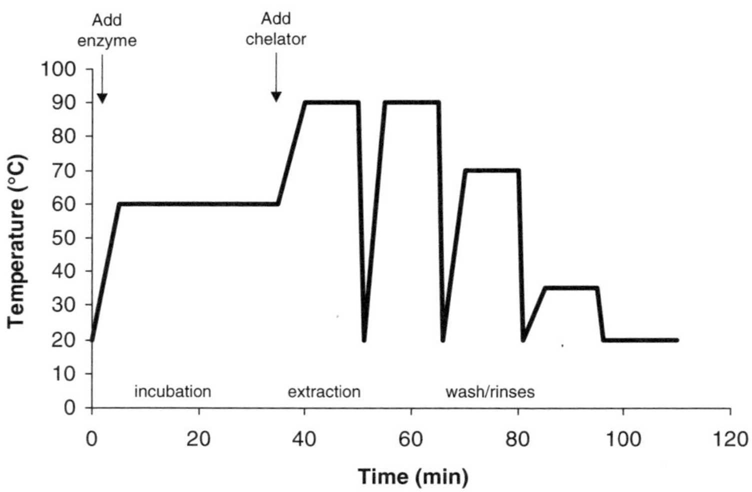

4.3 Typical time–temperature profile for a jet bioscouring process.

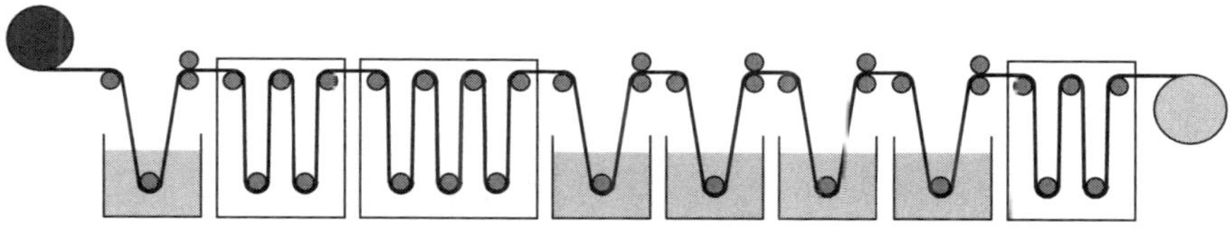

4.4 Process scheme for a (semi-)continuous pad-steam bioscouring process.

and the removal of waxes. A buffer is needed, e.g. a phosphate or citrate buffer, to maintain the pH between 7 and 9.5 (optimal pH 8.5–9.0). Combined with desizing the pH should not exceed 7.5–8.0 (see Section 4.3.1). The Ca^{2+} concentration is an important parameter in the enzymatic process, its presence slowing down the degradation of pectin but stabilising the enzyme. Therefore the addition of strong chelators is recommended only for the extraction and washing/rinsing phase. The weakened outer layers can be removed in the washing and rinsing process at a temperature above the melting point of the waxes (75–95°C), in the presence of chelators, emulsifiers and wetting agents. The process conditions for a pad-batch system are more or less identical, except that the incubation phase needs to be 1–4 hours at 60°C and 12–16 hours at 25°C.

In (semi-)continuous pad-steam machinery much shorter processing times can be realised (Lange and Henderson, 2000). The process scheme and a typical time–temperature profile for a continuous enzymatic scouring process are shown in Fig. 4.4 and 4.5, respectively. The processing conditions resemble those of the batch process described above except that the temperature during the incubation phase might be raised up to 95°C. Lower temperatures, but above 55°C, might be applied as well in this process. This is because the fabric being introduced does not reach full the temperature immediately and thus the enzyme has time to degrade the pectin before being deactivated by the high temperature. Other modifications to increase the process speed are a short hot-water treatment (Hartzell and Hsieh, 1998; Agrawal *et al.*, 2002) or a rinse at 50°C (Waddell, 2002) prior to the bioscouring process.

4.4 Mass transfer in textile materials

4.4.1 The structure of textile materials

Textile materials can have different structures such as woven and knitted fabrics, and non-wovens. In the context of this chapter we limit ourselves

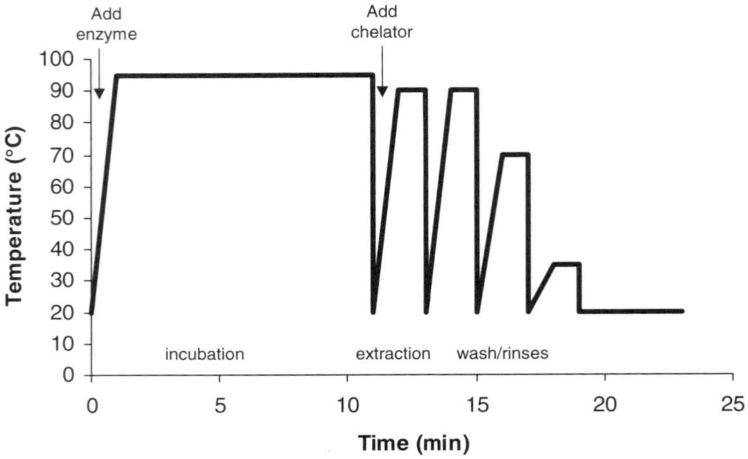

4.5 Typical time–temperature profile for a pad-steam bioscouring process.

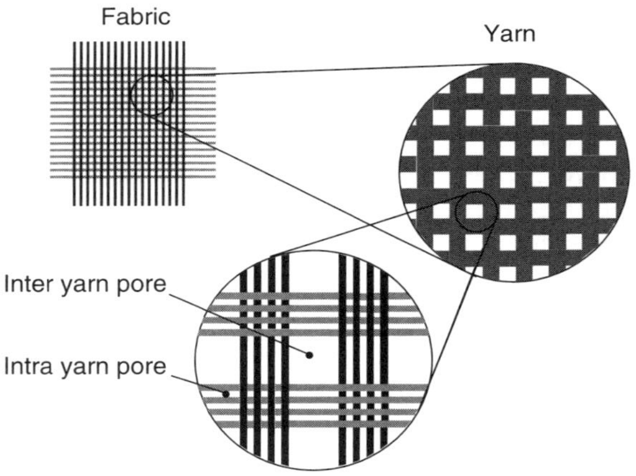

4.6 Structure of a woven textile material (schematically).

to discussing woven textiles. To make a woven textile, fibres are spun into yarns and yarns are woven into fabrics. This means that textiles can be seen as a porous slab with two kind of pores, pores between the fibres, the intra-yarn pores and pores between the yarns, the inter-yarn pores. This is schematically drawn in Fig. 4.6. This is why we say that woven textiles have a dual porosity. Figure 4.7 shows an example of the specific pore volume distribution in a cotton fabric, measured by a TRI-autoporosimeter (Textile

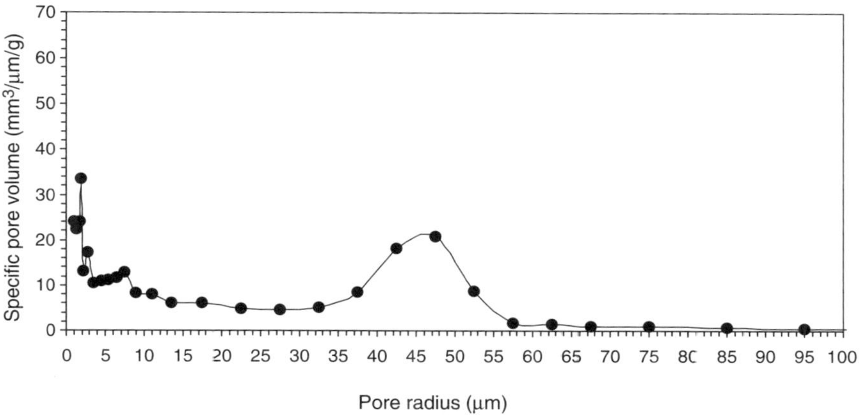

4.7 Pore size distribution of a woven cotton fabric.

Research Institute). From this figure it is clear that the cloth contains small pores, in the order of 2 μm, the intra-yarn pores, and larger pores in the order of 47 μm, the inter-yarn pores. Many investigators have studied and measured the pore size distributions in textile materials with respect to flow phenomena in textiles (Van den Brekel, 1987; Van den Brekel and de Jong, 1988; Gooijer, 1998).

The migration of the enzyme molecules into the intra-yarn pores is necessary for good enzymatic treatment of the fibres within a yarn. This can be achieved by flowing an enzyme solution through the fabric. However, since the flow resistance in the intra-yarn pores is much higher than the resistance in the inter-yarn pores, the bulk of the liquid will flow along the yarns instead of through the yarns. This was found by Van den Brekel and later confirmed by Gooijer. In Fig. 4.8 the flow pattern of a liquid flowing along a yarn is drawn schematically. Based on this Warmoeskerken and Boom (1999) introduced the concept of a stagnant core and a convective shell. The stagnant core of the yarn is the area in which there is no flow at all. The convective shell is the outer area of the yarn in which the flow penetrates to some extent. The transfer processes in the stagnant core are based on molecular diffusion while the transport processes in the outer convective shell are driven by convective diffusion. Since convective diffusion is much faster than molecular diffusion, the rate of mass transfer in the yarn will be determined by the size of the stagnant core. This means that the migration time of enzymes into the intra yarn pores is determined by molecular diffusion in the stagnant core, which is a relatively slow process.

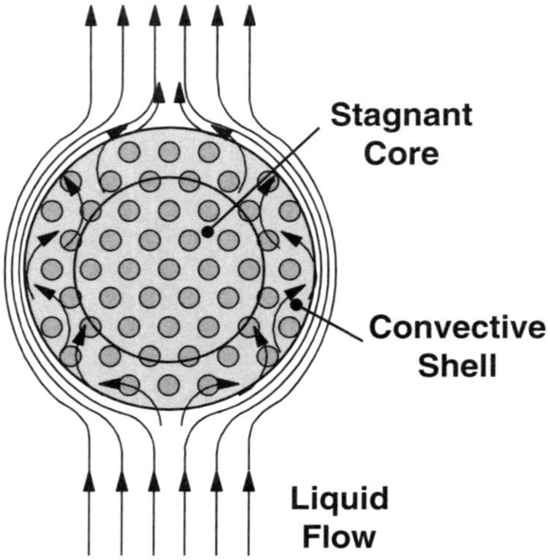

4.8 Liquid flow around and through a textile yarn. The dots represent the fibres in the yarn.

4.4.2 Diffusion of enzymes in a yarn

The diffusion time of enzymes in a yarn can be calculated by applying the theory for molecular diffusion that can be found in textbooks by Crank (1956) and Carslaw and Jaeger (1989). In the following approach, the direction of diffusion can be from the outside area to the centre of the yarn or from the yarn centre to the outside area. For the diffusion model we adapted the diffusion in a cylinder. The general equation describing this diffusion process in cylindrical coordinates is:

$$\frac{\partial C}{\partial t} = D\left[\frac{1}{r}\frac{\partial}{\partial r}\left(r\frac{\partial C}{\partial r}\right) + \frac{1}{r^2}\frac{\partial^2 C}{\partial \varphi^2} + \frac{\partial^2 C}{\partial z^2}\right] \quad [4.1]$$

in which t is the time, C is the time- and place-dependent concentration of the enzymes and r, φ, and z are the axes along which the diffusion process proceeds, see Fig. 4.9.

D is the diffusion coefficient in m^2/s of the diffusing enzyme. If only diffusion in the radial direction is considered, equation [4.1] reduces to:

$$\frac{\partial C}{\partial t} = \frac{D}{r}\frac{\partial}{\partial r}\left(r\frac{\partial C}{\partial r}\right) \quad [4.2]$$

This equation can be solved for different initial and boundary conditions. In the current case we consider the situation in which the enzymes diffuse from a bulk solution into the yarns and we assume that the enzyme concentration in the bulk remains constant because the volume of the liquid

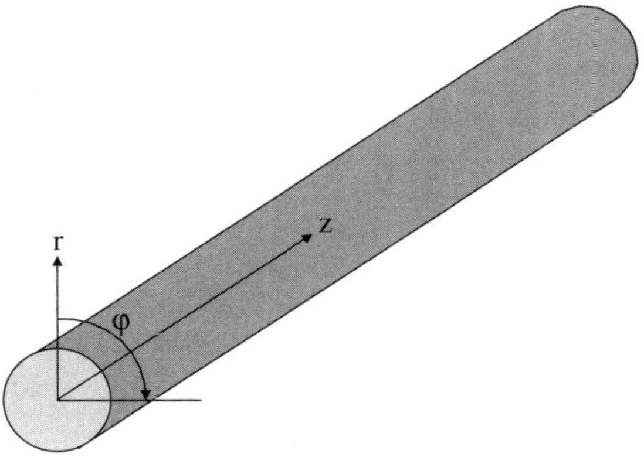

4.9 Schematic representation of the coordinate system used to model diffusion in a yarn.

bulk is much higher than the volume of the intra-yarn pores. This means that the concentration of the enzymes at the outer surface of the yarn is constant and equal to the bulk concentration. The second assumption is that at the start of the diffusion process no enzymes are present in the yarn. So the initial and boundary conditions can be written as:

$$t = 0 \quad 0 \leq r \leq \frac{1}{2} d_{\text{yarn}} \quad C = 0 \qquad [4.3]$$

$$t > 0 \quad r = \frac{1}{2} d_{\text{yarn}} \quad C = C_{\text{bulk}}$$

with C_{bulk} being the enzyme concentration in the bulk and d_{yarn} the yarn diameter.

The solution of equation [4.2] with the initial and boundary conditions according to equation [4.3] is a Bessel function and reads:

$$E = \frac{\overline{C}}{C_{\text{bulk}}} = 1 - \sum_{n=1}^{n=\infty} \left[\frac{4}{\mu_n^2} \exp(-4\mu_n^2 F_0) \right] \qquad [4.4]$$

with $\overline{C}$ being the average enzyme concentration in the yarn and:

$\mu_1 = 2.4048$
$\mu_2 = 5.5201$
$\mu_3 = 8.6537$
$\mu_4 = 11.7953$
$\mu_5 = \mu_4 + \pi$
. . .
. . .
. . .
$\mu_n = \mu_{n-1} + \pi$

F_0 is the dimensionless Fourier number and is defined as:

$$F_0 = \frac{Dt}{d_{\text{yarn}}^2} \qquad [4.5]$$

and the mean enzyme concentration in the bulk is defined as:

$$\overline{C} = \frac{2}{d_{\text{yarn}}} \int_0^{\frac{1}{2}d_{yarn}} C\mathrm{d}r \qquad [4.6]$$

The F_0 number represents the ratio of the process time t and the diffusion time d_{yarn}/D. E in equation [4.4] is the dimensionless mean concentration of the enzymes in the yarns. This is also called the efficiency of the diffusion process. At $t = 0$, when the mean enzyme concentration in the yarn is zero, the value of E is 0 as well, and after an infinitely long time when the mean enzyme concentration in the yarn equals the bulk concentration, E has the value of 1. Figure 4.10 shows the calculated result of equation [4.4].

In the calculations the value of n was taken as 25. However, the solution of the diffusion equation in the form of Bessel functions is not very easy to use. To overcome that problem the approximation method of Etters (1980) is very useful. He fitted the exact solution of equation [4.1] by:

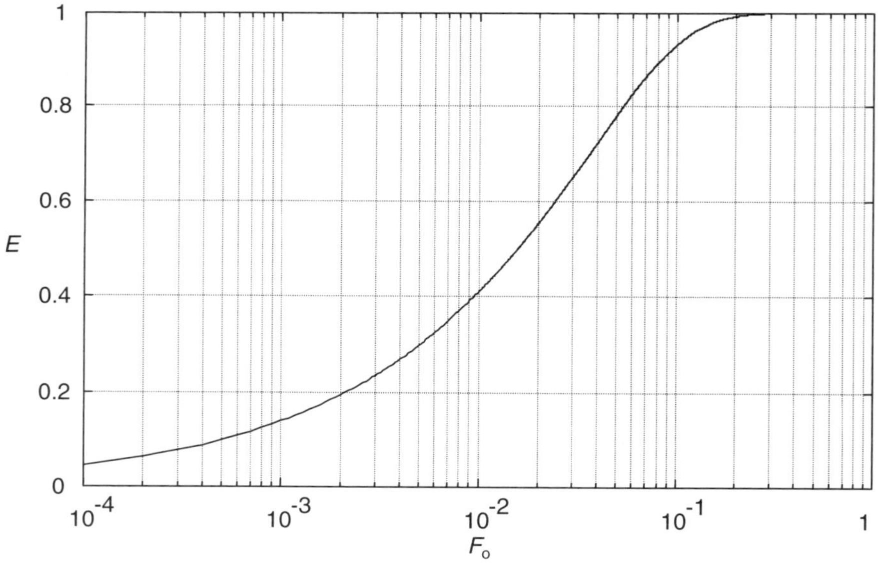

4.10 The exact solution of the diffusion problem (equation [4.4]).

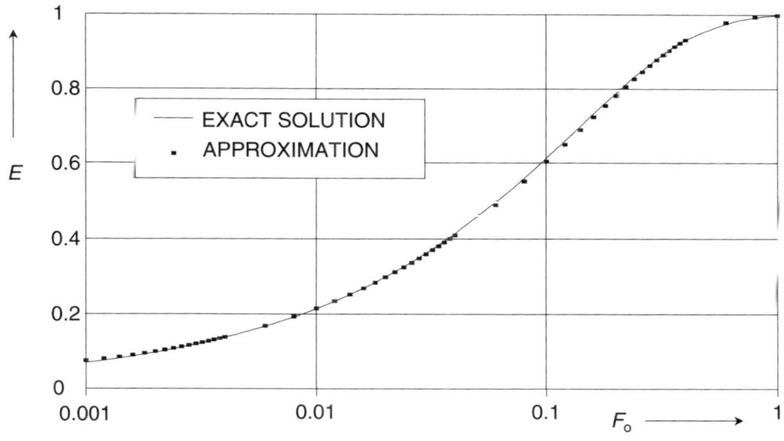

4.11 The approximate solution from Etters of the diffusion problem, and the exact solution.

$$E = \frac{\overline{C}}{C_{\text{bulk}}} = \left[1 - e^{-a(4F_0)^b}\right]^c \qquad [4.7]$$

In our case where there is no diffusion boundary layer between the bulk solution and the yarn, see the boundary conditions in equation [4.3], the values for the constants are $a = 5.530$, $b = 1.0279$ and $c = 0.3341$. Figure 4.11 shows the calculated results of E as a function of F_0 according to the exact solution, equation [4.4], and according to the approximation formula, equation [4.7]. From this figure it can be concluded that the approximation formula of Etters leads to good results. With this formula the diffusion time can be calculated if the diffusion coefficient of the enzymes is known.

4.4.3 The diffusivity of enzymes

Since we have worked with dimensionless numbers like the Fourier number, until now we did not need the value for the diffusion coefficient of enzyme molecules. If we want to make calculations for the diffusion time of enzymes in yarns we need to find values for the diffusion coefficients. Measurement of the diffusion coefficient is rather complicated. Most methods, mentioned by Van Holde (1971) and by Sun (1994), are based on the application of Fick's first law for diffusion. This law reads:

$$V\frac{\mathrm{d}C}{\mathrm{d}t} = DA\Delta C \qquad [4.8]$$

in which V is the liquid volume in m^3 in which the diffusion process proceeds, A is the surface area in m^2 through which the molecules diffuse, D is the diffusion coefficient in m^2/s, C is the concentration of the diffusing component in kg/m^3 and ΔC is the concentration difference. The most common procedure is to separate two liquids from each other by a membrane. In one liquid the diffusing component is absent at time $t = 0$. In the other liquid the diffusion of the component is followed by optical methods such as Schlieren and refraction.

It is also possible to calculate the diffusion coefficient; we can apply the Stokes–Einstein relation (Bird *et al.*, 1960). This relation reads:

$$D = \frac{k_B T}{3\pi\eta_L d_M} \qquad [4.9]$$

in which D is the diffusion coefficient in m^2/s, k_B is the Boltzman constant in J/K, T the temperature in K, η_L the dynamic viscosity of the liquid in which the molecules diffuse in Pa.s, and d_M the diameter of the diffusing molecule in m. Assuming the enzyme molecule is a sphere, its diameter can be calculated by:

$$d_M = \sqrt[3]{\frac{6M_W 10^{-3}}{N_{AV}\rho_M\pi}} \qquad [4.10]$$

in which M_W is the molecular weight of the enzymes in kg/kmol, N_{AV} is the Avogadro number in mol^{-1} and ρ_M is the density of the molecule in kg/m^3.

From equations [4.9] and [4.10] follows the relation between the molecular weight of the enzyme molecule and its diffusivity in a liquid:

$$D = \frac{k_B T}{3\pi\eta_L \sqrt[3]{\frac{6M_W 10^{-3}}{N_{AV}\rho_M\pi}}} \qquad [4.11]$$

Figure 4.12 shows some results of this equation. In this figure the measured data of Daniels and Alberty (1975) for the diffusion coefficient of large protein molecules in water are compared with the calculated values according to equation [4.11]. From the figure it can be concluded that equation [4.11] gives reasonable to good results for the diffusion coefficients of enzymes.

Since we are focusing on the diffusion of enzymes in textiles we have to take account of the porosity of the system. The effect of the porosity on the diffusion coefficient can be expressed as:

$$D_{porous} = D\varepsilon \qquad [4.12]$$

in which D_{porous} is the effective diffusion coefficient and ε the porosity of the system. So the diffusion coefficient in porous systems is smaller than that in homogeneous systems. This is because in the yarn the fibres decrease

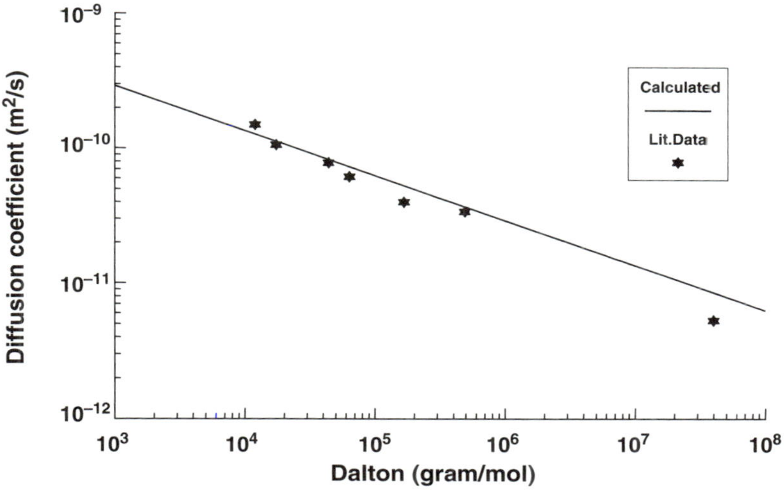

4.12 Calculated and measured diffusion coefficients (experimental data from Daniels and Alberty, 1975).

the free area through which the enzymes diffuse. Another aspect that affects diffusion in porous systems is the tortuosity. In a yarn the enzyme molecules cannot diffuse via the shortest path, that is via the radius to the centre, because the fibres obstruct the straight path to the centre. So in reality the enzymes have to diffuse through a labyrinth of fibres, which makes the actual diffusion path longer. This effect is expressed in the tortuosity β, which gives the ratio between the actual path length for diffusion and the free path length. For a yarn consisting of fibres, the tortuosity β has a value of 2. This means that the actual diffusion length in the radial direction of the yarn is twice the radius of the yarn. From the diffusion equation it can be derived that the diffusion coefficient in a tortuous system is:

$$D_{\text{tortuous}} = \frac{D}{\beta^2} \qquad [4.13]$$

If we now combine equations [4.12] and [4.13] we find an expression for the effective diffusion coefficient D_{eff} that includes the effect of porosity as well as that of tortuosity:

$$D_{\text{eff}} = \frac{D\varepsilon}{\beta^2} \qquad [4.14]$$

From this equation we can calculate that the diffusion coefficient of enzymes in a yarn with a porosity $\varepsilon = 0.5$ and a tortuosity $\beta = 2$ is a factor of 0.125 smaller than the diffusion coefficient in a homogeneous system.

In an example we will now calculate the diffusion time of enzymes in a yarn. Suppose we have a yarn with a diameter of 0.5 mm, a porosity of 0.4 and a tortuosity of 2. In Fig. 4.10 we see that the diffusion process is more or less completed when the F_o number has a value of 1. With equation [4.5] we can derive:

$$t = \frac{F_o d_{yarn}^2}{D_{eff}} \qquad [4.15]$$

or with the values for F_o and d_{yarn} chosen above:

$$t = \frac{2.5 \times 10^{-7}}{D_{eff}} \qquad [4.16]$$

The effective diffusion coefficient as function of the molecular weight of enzyme molecules can be calculated by equations [4.11] and [4.14]. We have done this for the following values of the parameters:

ρ_M	=	1000	kg/m^3
μ_L	=	10^{-3}	Pa.s
T	=	293	K
N_{AV}	=	6×10^{23}	mol^{-1}
k_B	=	1.38×10^{-23}	J/K

Figure 4.13 shows the results of the calculations. From this figure it is clear that the time to complete the diffusion process is already in the order of 5 hours for an enzyme with a molecular weight of 20 000 g/mol. This process time is not available in continuous textile treatment processes.

We have repeated the calculations for different materials such as dye molecules, enzymes, carbon black particles and silica particles. In each

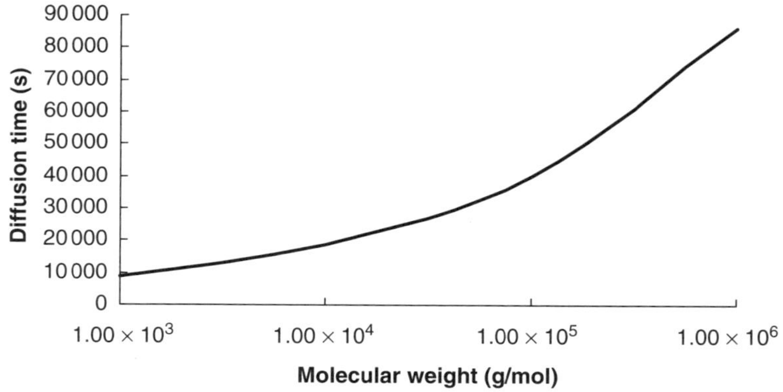

4.13 Time needed to complete the diffusion process.

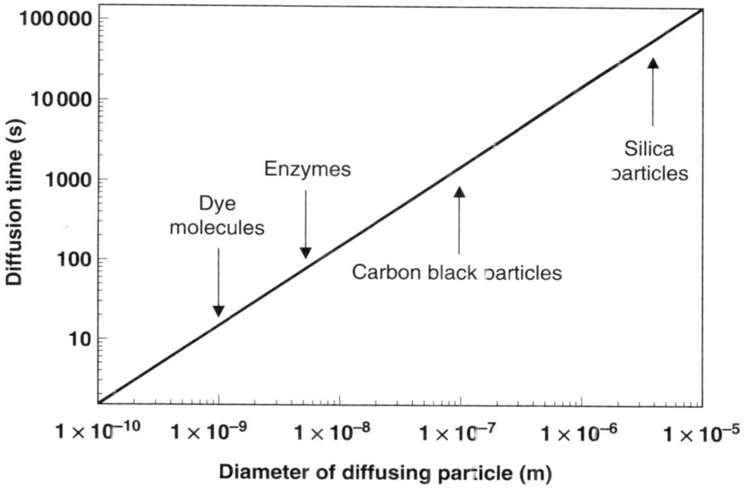

4.14 Time needed to remove 90% of particles from a yarn by diffusion as a function of the particle diameter.

case we calculated the time needed to complete 90% of the diffusion process in a yarn. In the case $F_o = 0.1$, see Fig. 4.10, the results have been drawn in Fig. 4.14 in which the calculated diffusion time is plotted against the diameter of the diffusing particle. From the figure it is clear that the diffusion times are lower for lower values of F_o, although they are still high compared to process times. Figure 4.14 also shows that the diffusion time increases non-linearly with the diameter of the diffusing particle.

Thus, taking into account the typical porous structure of textile materials and the relatively large size of the enzyme molecules, the physical transport of the molecules is often a rate-limiting step in enzymatic textile treatment processes. The only way to lower the transport time is by decreasing the stagnant core, thus creating convective flow in the intra-yarn pores of the fabric.

It has to be mentioned that this is only the case in so-called wet-to-wet applications. In wet-to-dry applications the enzyme solution can penetrate into the pores by the so-called wicking effect. This is the result of capillary forces which allow the enzyme solution to penetrate directly into the pores of the yarn. Thus it is clear that the transport of enzymes into the textile material is much faster in the case of wet-to-dry applications than in wet-to-wet systems owing to wetting and wicking. Wetting and wicking of textile materials is a subject beyond the scope of this chapter.

4.5 Process intensification: enhancement of mass transfer in textile materials

Process intensification is important when introducing new processes into the textile industry. Process intensification will not only result in more efficient and economically feasible processes, it also offers possibilities for production on demand because of a dramatic decrease in the residence time. From the data presented in Section 4.3 it is clear that the residence time in enzymatic textile pretreatment and finishing processes, like that in conventional textile processes is still relatively long. In Section 4.4 it was shown that the activity of enzymes in textile treatment processes can be limited to a large extend by their slow diffusion into the pores of a yarn. Decreasing the diffusional core or the stagnant core in the yarn can enhance this transport rate. In other words by creating flow in the intra-yarn pores, the rate of the mass transfer process is then determined by convection which is always much faster than diffusion. In the present section the possibilities of enhancing mass transfer through the deformation of textile materials and the application of ultrasound are discussed as tools for intensifying enzymatic wet textile processes.

4.5.1 Deformation of textile materials

The most common way to enhance mass transfer in the intra-yarn pores is to deform the porous matrix of the textile material. When a force is applied to the textile the pores become smaller resulting in a flow of the pore liquid to the treatment bath. If thereafter the force is released, the textile system relaxes, the pores recover their original shape and liquid flows from the bath into the pores. This so-called squeezing effect can be obtained in different ways. In an open-width process this phenomenon occurs at the moment the textile passes a roller. This is drawn schematically in Fig. 4.15.

On the side where the textile is attached to the roller the textile is deformed by compression forces. On the outside the textile is deformed by stretching forces. Although this phenomenon is well known, quantitative data about the internal flow in the textile pores that is generated by this mechanism is not known. At the point at which the fabric contacts the roller, the liquid is forced through the textile as indicated in Fig. 4.15. Also no quantitative data are available about this mechanism. This is why the design of this kind of process equipment is often based on trial and error. As far as we know, only Van der Donck *et al.* (1998) have done some work on the influence of squeezing on mass transfer. Farber and co-workers (Farber and Dahmen, 1998; Farber *et al.* 1999), for example, used advanced computational fluid dynamics (CFD) to describe mass transfer in open width equipment, but despite the quality of their model and the outcome of their

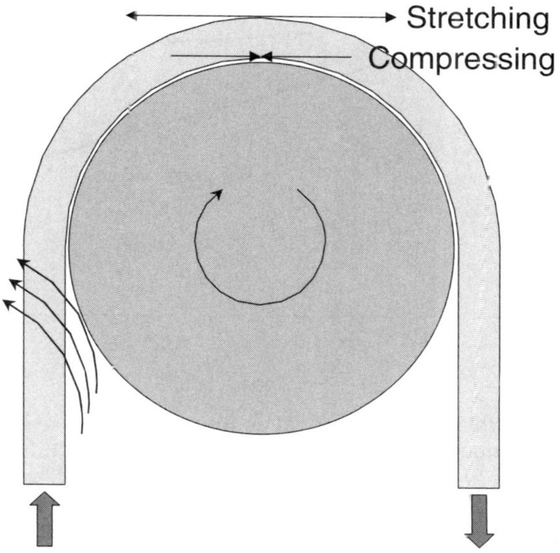

4.15 Deformation of a textile on a roller.

simulations, the textile material is still described as a homogeneous permeable rigid structure instead of a deformable biporous structure. Van der Donck *et al.* studied the squeezing effect in yarns when they are stretched and concluded that this mechanism contributes to a large extent to the mass transfer rate in textile yarns.

Van der Donck *et al.* reported that deforming yarns by placing a fabric in a pulsating flow or repeated deformation through mechanical elongation of the yarns improved mass transport compared with diffusion alone. When a yarn is elongated a quantity of liquid, as well as the enzymes or chemicals in that liquid, is squeezed out of the yarn. Van der Donck measured the increase of conductivity caused by the release of magnesium sulphate with time from a cotton yarn impregnated with magnesium sulphate. The magnesium sulphate was squeezed out of the yarn by the repeated elongation of that yarn. In practice this is realised at the rollers of open-width equipment or through tumbling the cloth in domestic laundry machines. To describe the phenomena that are observed, Van der Donck used the dimensionless Fourier number, which gave a qualitative logarithmic relation between the soil release, the Fourier number and the additional salt removed. However, Etters (1980) proposed a mathematically simple empirical equation that matches the exact solution to describe diffusion-controlled mass transport in yarns. Equation [4.7] has been extended in such

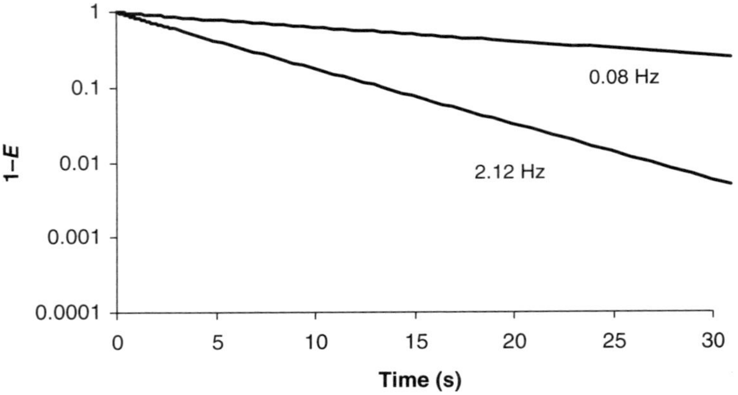

4.16 Calculated relative soil removal as function of time using the modified equation of Etters for different deformation frequencies. ($D_{eff} = 4.8 \times 10^{-10}\,m^2/s$, $d_{yarn} = 3.5 \times 10^{-4}\,m$, $\varepsilon_{intra\text{-}yarn} = 0.5$, $\varepsilon_{deformation} = 0.03$, $a = 2.440$, $b = 1.045$, $c = 0.863$). Van der Donck *et al.* used a frequency of 0.08 and 2.12 in their experiments.

a way that it is possible to describe the influence of the deformation of the yarn on mass transport:

$$1 - E = \left[1 - \exp\left(-a\left(\frac{4D_{eff}t}{d_{yarn}^2}\right)^b\right)\right]^c \left(\frac{\varepsilon_{intra\text{-}yarn} - \varepsilon_{deformation}}{\varepsilon_{intra\text{-}yarn}}\right)^{ft} \quad [4.17]$$

where $\varepsilon_{intra\text{-}yarn}$ is the porosity of the yarn, $\varepsilon_{deformation}$ is the volume fraction squeezed out of the yarn during deformation and f is the frequency of the deformation. With this modified equation of Etters we are able to describe increased mass transport caused by stretching the yarns. In domestic laundry processes the deformation of the plug and therefore of the yarns is related to the rotation velocity of the drum. Thus, in industrial open-width equipment and in a domestic laundry processes a deformation frequency between 0 and 2.5 Hz seems realistic. In Fig. 4.16 the results are shown using the modified equation from Etters. Different constants for a, b and c in Equation [4.17] were used to correct for the low amount of mixing in the experiments of Van der Donck *et al.* The calculated soil removal time compares very well with the results described by Van der Donck.

4.5.2 Ultrasound-enhanced mass transfer

Another way of enhancing the mass transfer is the application of ultrasonic waves. Ultrasound as a means of intensification of wet textile processes has been attempted by several researchers (e.g. McCall *et al.*, 1998; Thakore,

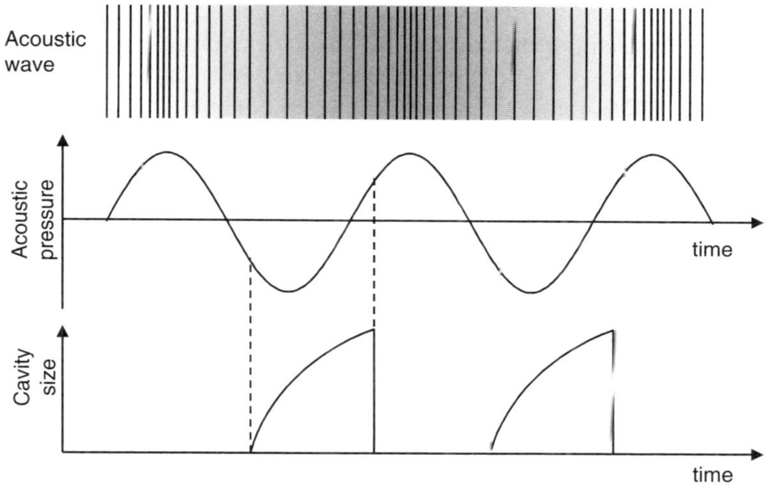

4.17 Some characteristics of an ultrasonic wave.

1990; Yachmenev *et al.*, 1998, 1999, 2001; Rathi *et al.* 1997). In spite of encouraging results in laboratory scale studies, ultrasound-assisted wet textile processes have not yet been implemented on an industrial scale. Two major factors that have contributed to this are lack of precise knowledge about the physical mechanism of ultrasonic mass transfer enhancement in textiles and the inherent drawbacks of ultrasonic processors, such as directional sensitivity, erosion of sonicator surface and the non-uniform volumetric energy dissipation.

Ultrasound is a longitudinal pressure wave in the frequency range above 25 kHz (see Fig. 4.17). As the sound wave passes through water in the form of compression and rarefaction cycles the average distance between the water molecules varies. If the pressure amplitude of the sound is sufficiently large, the distance between the adjacent molecules can exceed the critical molecular distance during the rarefaction cycle. At that moment a new liquid surface is created in the form of voids. This phenomenon is called acoustic cavitation and is drawn schematically in Fig. 4.17.

The theoretical pressure amplitude that causes cavitation in water is approximately 1500 bar. However, in practice acoustic cavitation occurs at a far lower pressure amplitude, less then 5 bar. This is due to the presence of weak spots in the liquid in the form of tiny microbubbles that lower the tensile strength of the liquid. Once formed, the bubbles can redisolve into the liquid; they may float away, or, depending on their size, they may grow and shrink in phase with the oscillating ultrasonic field. This process of growing and recompressing bubbles is called stable cavitation. If the sound field is sufficiently intense, bubbles of a specific initial size can grow so quickly and acquire such momentum that the compression wave that immediately follows the rarefaction phase is no longer able to stop the bubbles growing. Once out of phase with the ultrasonic field, however, the bubbles are no longer stable. The pressure within the bubble is not high enough to sustain the size of the bubble and, driven by the next compression wave, the bubbles implode. This latter process is called transient cavitation.

In liquids the collapsing bubbles remain spherical because the ultrasonic waves are uniform. However, if a transient acoustic bubble collapses near a solid boundary, the bubble will implode asymmetrically, generating jets of liquid directed towards the surface of the solid boundary. The microjets resulting from collapsing bubbles at a solid boundary account for the well-known cleaning effect of ultrasonic waves. This acoustic cavitation process has been described by many authors, for example Neppiras (1980), Apfel (1981) and Suslick (1988).

This area of power ultrasound is what we have called sonomechanics, making a clear distinction from sonochemistry. The latter is the application of power ultrasound to speed up chemical reactions. The idea is that if an acoustic cavity collapses adiabatically the temperature of the very small volume of liquid involved must rise several thousands of degrees K enhancing the rate of chemical reaction. A lot of literature about this subject is available, for example Mason and Lorimer (1989) and Mason (1990). Ultrasound has also been found to have an effect on enzymatic reactions (Warmoeskerken *et al.*, 1994). However it is not clear whether the observed enhancement in the enzymatic reaction rate is due to the temperature effect mentioned above or to a more intrinsic effect, such as unfolding the enzyme molecule so that the reactive site becomes more accessible to the substrate. More fundamental research is needed in this area to clarify the mechanism and to develop a process in which the enzymatic reactions are boosted by ultrasound.

Here we restrict ourselves to sonomechanics, the application of ultrasound to enhance mass transfer in textile materials. Much research can also be found in this area (Yachmenev *et al.* 1998, 1999, 2001; Moholkar 2002; Moholkar and Warmoeskerken, 2000, 2001, 2002; Moholkar and Pandit, 2001; Moholkar *et al.*, 2000, 2002; Warmoeskerken *et al.*, 2002).

We have found that ultrasound can speed up mass transfer in textile materials. Figure 4.18 shows the results of a typical experiment (Warmoeskerken, 2002). A textile cloth was impregnated with a salt that was then washed out in water. The release of salt from the textile with time was followed by conductivity measurements in the bath. One experiment was performed without ultrasound and one with ultrasound. The results of these experiments are shown in Fig. 4.18 where $(1 - E)$, representing the fraction of salt that is still on the cloth, is plotted against the process time.

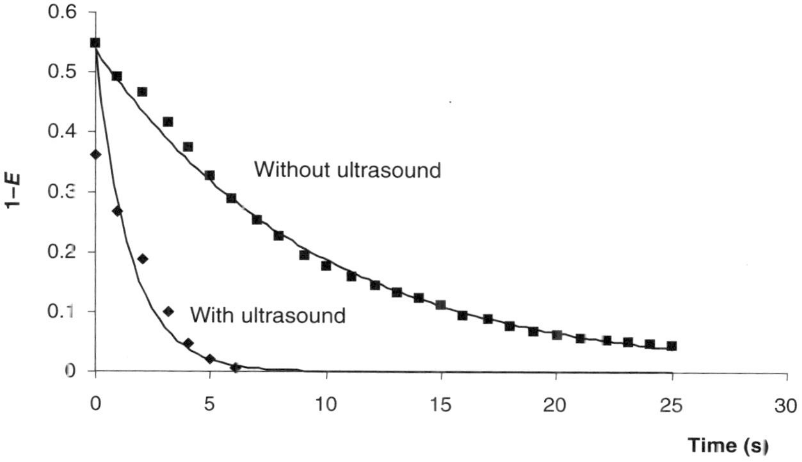

4.18 Experimental salt-rinsing results with and without the application of ultrasound.

From Fig. 4.18 it is clear that in the case where ultrasound is applied salt release is much faster than in the case without ultrasound.

The mechanism of this phenomenon involves the formation of transient acoustic cavities in the close vicinity of the textile surface. These asymmetrically collapsing cavities create locally microscale liquid jets that are directed towards the substrate surface and penetrate deeply into the pores of the textile. It can be argued, with respect to the stagnant core–convective shell model, that ultrasound decreases the stagnant core in the yarn, resulting in an enhanced mass transfer rate in the yarn.

The development of a more efficient textile treatment or enzymatic textile treatment process based on these findings is not simple. An ultrasonic system is quite complex and the performance is dependent on all the system parameters, for example the size of the system, the properties of the ultrasonic equipment, the frequency and power of the ultrasonic wave and the composition of the liquid. A very important parameter seems to be the presence of air in the water. The ultrasonic wave, while traveling from the transducer to the substrate will create a lot of acoustic cavities in the bulk liquid between the transducer and the substrate. However, since the transient cavities are only required at the substrate surface, the formation of cavities in the bulk liquid is in fact only a waste of energy. In a deaerated system there are no energy losses during the time that the ultrasonic wave travels from the transducer to the substrate and all the ultrasonic energy is applied to the formation of transient cavities at the substrate surface. Therefore some air pockets have to be available at the substrate surface. In

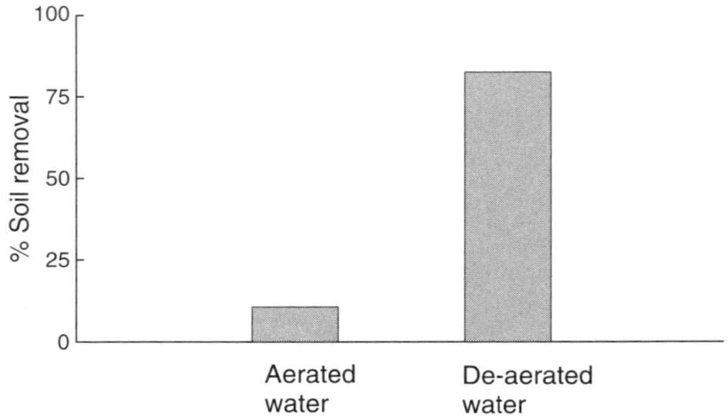

4.19 Soil removal from Empa 101 with the application of ultrasound in aerated and de-aerated water.

practice there is always sufficient air present in the textile material for the intended ultrasonic effect.

Figure 4.19 shows some results of cleaning an Empa 101 test cloth (Moholkar, 2002). This is a test monitor that is used to study the performance of laundry systems and is cotton impregnated with a mixture of olive oil and carbon black particles. Figure 4.19 shows that in the deaerated case the performance of the ultrasonic wave, in terms of soil removal, is more than seven times better than in the aerated case. Although these are very promising results, a lot of research is still needed to translate the current knowledge about ultrasonically boosted wet textile treatment processes to an operational full-scale process.

4.6 Mass transfer and diffusion limitation in immobilised enzyme systems

In some industrial textile processes enzymes or microorganisms are immobilised; for example to prevent them from flushing out of the (bio)reactor during a continuous operation such as the decolourisation of textile effluents (Oxspring *et al.*, 1996; Mielgo *et al.*, 2001, 2002). To immobilise biocatalysts (enzymes or complete microorganisms) several methods are available such as entrapment in a porous support or attachment of the biocatalyst to a surface (for more details see e.g. Shuler and Kargi, 2002 and van't Riet and Tramper, 1991). In general the most important advantages of immobilising biocatalysts are:

- the possibility of reuse;
- the possibility of continuous operation;
- a more stable or active enzyme (different microenvironment);
- elimination of enzyme recovery and purification processes.

However, immobilisation of enzymes or microorganisms can cause mass transfer limitations compared to free enzymes or microorganims. Diffusional resistance may be observed at different levels depending on the nature of the support, the turbulence in the reactor (the Reynolds number), the particle size and the distribution and concentration of the enzyme over the particle. Simply decreasing the particle size and increasing the porosity to eliminate or to reduce mass transfer problems is not always desirable for optimal reactor performance (Van 't Riet and Tramper, 1991).

Whether diffusion resistance in a heterogeneous biocatalyst (the particle, or 'bead' with immobilised enzymes or microorganisms) has a significant effect on the overall reaction rate depends on the ratio of the enzymatic reaction rate and the diffusion rate, which is characterised by the Damköhler number (Da).

$$Da = \frac{\text{maximum rate of reaction}}{\text{maximum rate of diffusion}} = \frac{V_{max}}{k_{sl} A' K_m} \qquad [4.18]$$

where V_{max} (mol/(s·m^3)) is the maximal rate of conversion, K_m (mol/m^3) is the Michaelis–Menten constant, k_{sl} is the mass transfer coefficient (m/s) and A' is the specific surface area (m^{-1}). If $Da >> 1$, the diffusion rate in the heterogeneous biocatalyst is limiting; for $Da << 1$, the reaction rate is limiting compared to mass transfer by diffusion; and for $Da \approx 1$, the diffusion and reaction rates are comparable.

4.6.1 Diffusion effects in immobilised enzyme systems

Mass transfer limitation can occur in the biocatalyst particle itself, in the external layer surrounding the particle, or in both (see Fig. 4.20). In the bead, mass transfer is in principle only possible by diffusion. Concentration gradients are the driving force for diffusion. Because of the consumption of substrate by enzyme, the concentration in the particle will decrease and a concentration gradient will develop. This results in a net flow of substrate from the surface of the particle to the centre of the particle. Diffusion limitation in the particle is thus a function of, for example, the particle size, the porosity and the amount of enzyme in the particle.

A thin stagnant liquid film surrounding the particle causes mass transfer limitation by diffusion in the external layer. The thickness (δ) of that stagnant film, and thus the amount of external mass transfer limitation, depends on the flow conditions in the reactor and can be described by film theory. The same analysis of external and internal diffusion limitation holds for biofilms and microbial flocs (aggregates of cells).

In general the substrate concentration in the biocatalyst particle and in the stagnant film surrounding the particle will be lower than the concentration in the continuous or bulk phase. This means that if the biocatalyst obeys Michaelis–Menten kinetics (see the preceeding chapter), the

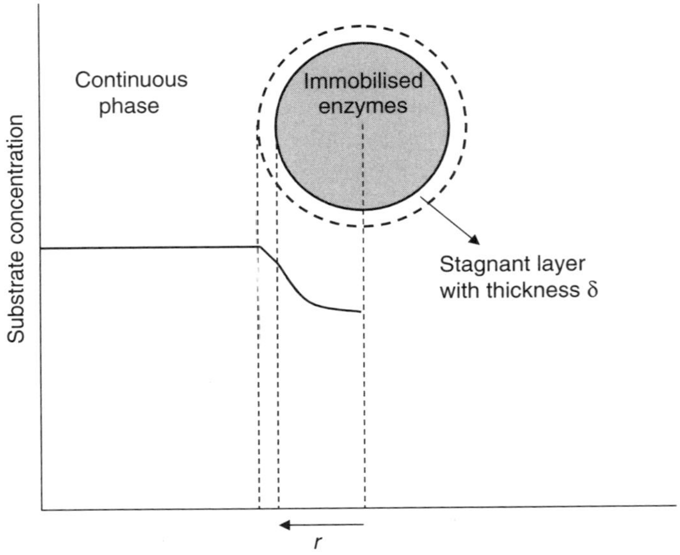

4.20 Substrate concentration profile in and around an immobilised enzyme particle (the continuous phase is perfectly mixed).

observed reaction rate will be lower than could be expected on the basis of the bulk concentration. Quite often, apparent or lumped kinetic constants are used to describe the kinetics of immobilised biocatalysts:

$$-r_s = -\frac{\partial C_{s,bulk}}{\partial t} = V_{max}^{app} \frac{C_{s,bulk}}{K_m^{app} + C_{s,bulk}} \quad [4.19]$$

in which r_s is the rate of substrate conversion (mol/(s·m^3)), t is the time (s), $C_{s,bulk}$ is the substrate concentration in the bulk liquid (mol/m^3), V_{max}^{app} (mol/(s·m^3)) is the apparent maximal rate of conversion and K_m^{app} (mol/m^3) is the apparent Michaelis–Menten constant. However, these apparent constants depend on the radius of the particle R_p, the effective diffusion coefficient D_{eff}, and the true reaction constants K_m and V_{max}. Therefore this approach is incorrect in principal. It is better to introduce, as in heterogeneous catalysis, an overall effectiveness factor η^{ov} (–) (Van 't Riet and Tramper, 1991):

$$-r_s = -\frac{\partial C_{s,bulk}}{\partial t} = \eta^{ov} V_{max} \frac{C_{s,bulk}}{K_m + C_{s,bulk}} \quad [4.20]$$

The overall effectiveness factor is defined as the product of the external and the internal effectiveness factor:

$$\eta^{ov} = \eta_e \eta_i \quad [4.21]$$

The overall effectiveness factor usually lies between 0 and 1 (Michaelis–Menten kinetics) and depends on the substrate concentration. It may sometimes be higher than unity owing to non-isothermal operation, partition effects or more complicated kinetics such as substrate inhibition kinetics.

4.6.1.1 External diffusion limitation

As explained above, because there is a thin stagnant film surrounding the biocatalyst particle, the substrate concentration is lower at the surface compared with the bulk concentration and thus the enzymes that are immobilised at the surface of the particle experience a lower substrate concentration. This is also the case for surface-bound enzymes on non-porous support materials (Fig. 4.21).

In order to correct for this an external effectiveness factor, η_e (–), is introduced, as in heterogeneous catalysis. This external effectiveness factor is defined as the ratio of the reaction rate at $C_s = C_{s,surf}$ ($C_{s,surf}$ is the substrate at the surface of the particle) and the reaction rate if external diffusion limitation is absent $C_s = C_{s,bulk}$:

$$\eta_e = \frac{-r_s(C_{s,surf})}{-r_s(C_{s,bulk})} \qquad [4.22]$$

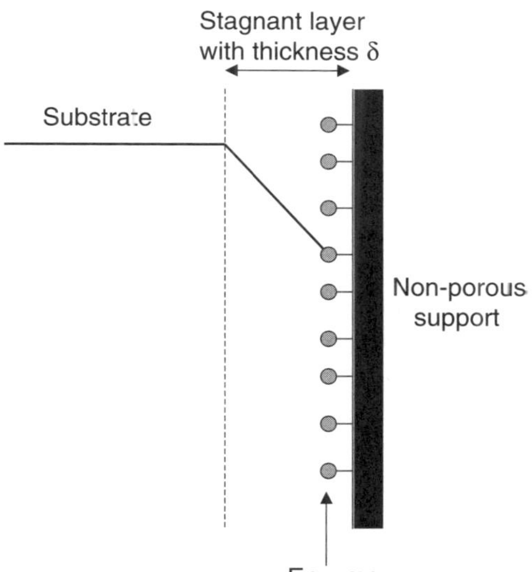

4.21 Substrate concentration profile in a stagnant layer with enzymes bound to the surface of a non-porous support.

For Michaelis–Menten kinetics this results in:

$$\eta_e = \frac{C_{s,surf}(K_m + C_{s,bulk})}{C_{s,bulk}(K_m + C_{s,surf})} \quad [4.23]$$

We assume that the substrate concentrations in the bulk liquid and K_m are known. This means that in order to determine the external effectiveness factor the substrate concentration at the surface should be known. In the steady state, the reaction rate at the surface is equal to the mass transfer rate J_s:

$$J_s = -r_s(C_{s,surf}) = \frac{V_{max} C_{s,surf}}{K_m + C_{s,surf}} = k_{sl} A'(C_{s,bulk} - C_{s,surf}) \quad [4.24]$$

k_{sl} and r_s can be measured and therefore $C_{s,surf}$ and the external effectiveness factor can be calculated. Equation [4.23] can be simplified for extreme values. If the substrate concentration at the surface is much greater than K_m, equation [4.24] reduces to:

$$J_s = -r_s(C_{s,surf}) = V_{max} = k_{sl} A'(C_{s,bulk} - C_{s,surf}) \quad [4.25]$$

Often, however, $C_{s,surf} << K_m$. Under such conditions we can consider two limiting cases. For $Da >> 1$, we can write:

$$-r_s(C_{s,surf}) = k_{sl} A' C_{s,bulk} \quad [4.26]$$

For $Da << 1$, we can write:

$$-r_s(C_{s,surf}) = C_{s,bulk} \frac{V_{max}}{K_m} \quad [4.27]$$

4.6.1.2 Internal diffusion limitation

When biocatalysts are immobilised on the internal pore surfaces of a porous matrix, substrate diffuses through a tortuous pathway among pores and reacts with the enzyme immobilised on the pore surfaces. Diffusion and reaction occur simultaneously (see Fig. 4.20) and consequently a radial concentration gradient will develop in the particle in addition to the gradient in the thin stagnant film surrounding the particle. Therefore the reaction rate (assuming Michaelis–Menten kinetics and no partition effects) will be lower as well. This phenomenon is accounted for by the internal effectiveness factor, η_i (–). The internal effectiveness factor can be described as the observed reaction rate divided by the rate which would be observed if all biocatalyst experienced a substrate concentration equal to that at the surface of the porous biocatalyst bead:

$$\eta_i = \frac{-r_s}{-r_s(C_{s,surf})} = \frac{4\pi R_p^2 D_{eff} \left(\frac{\partial C_s(r)}{\partial r} \right)_{r=R_p}}{-r_s(C_{s,surf})} \quad [4.28]$$

$\partial C_s(r)/\partial r$ at the surface of the particle is determined by the kinetics of the enzyme. Assuming Michaelis–Menten kinetics, no partitioning effects, and that the enzyme is distributed homogeneously over the porous support, we can write in the case of a steady state:

$$D_{\text{eff}}\left(\frac{1}{r^2}\frac{\partial}{\partial r}\left(r^2\frac{\partial C_s}{\partial r}\right)\right)=\frac{V'_{\max}C_s}{K_m+C_s} \qquad [4.29]$$

where V'_{max} is V_{max} per m^3 immobilisation material. Equation [4.29] can be solved numerically using the appropriate boundary conditions ($C_s = C_{s,surf}$ at $r = R_p$; and $\partial C_s(r)/\partial r = 0$ at $r = 0$), in order to determine $\partial C_s(r)/\partial r$ at the surface.

In order to quantify the internal effectiveness factor, the Thiele modulus (Φ) is often used (Aris, 1965; Bischoff, 1965). The Thiele modulus is the ratio of the kinetic rate and the diffusion rate:

$$\Phi=\frac{R_p}{3}\frac{-r_s(C_{s,\text{surf}})}{\sqrt{2}}\left[\int_{C_s^*}^{C_{s,\text{surf}}} D_{\text{eff}}(-r_s)\,\mathrm{d}C_s\right]^{-0.5} \qquad [4.30]$$

C^* is the concentration when thermodynamic equilibrium is reached. In case of Michaelis–Menten kinetics, C^* equals zero. For Michaelis–Menten kinetics the Thiele modulus is defined as:

$$\Phi=\frac{R_p}{3}\sqrt{\frac{\dfrac{V'_{\max}C_{s,\text{surf}}}{K_m+C_{s,\text{surf}}}}{2D_{\text{eff}}(K_m+C_{s,\text{surf}})\left(1+\dfrac{K_m}{C_{s,\text{surf}}}\ln\left(\dfrac{K_m}{K_m+C_{s,\text{surf}}}\right)\right)}} \qquad [4.31]$$

For zero-order and first-order kinetics ($C_s >> K_m$ and $C_s << K_m$, respectively) analytical solutions do exist (for details see Van 't Riet and Tramper, 1991). A plot of η_i versus the Thiele modulus results in a generalised figure (see e.g. Aris, 1965; Bischoff, 1965; Van't Riet and Tramper, 1991). In scientific literature simplified expressions of the Thiele modulus are used as well (Shuler and Kargi, 2002); however these expressions do not lead to a generalised relationship between the Thiele modulus and the internal effectiveness factor.

Internal diffusion limitation does not only affect the transport of substrate into the immobilised biocatalyst particle, but also the transport of the product(s) formed by the biocatalyst to the bulk liquid. As a consequence of the formation of products in the particle, there will be a net flow of product(s) from the centre of the particle towards the surface of the immobilised biocatalyst particle. A radial concentration gradient will develop in the particle and additionally a concentration gradient will develop in the stagnant liquid film surrounding the particle. The average product concentration in the microenvironment of the biocatalyst is thus higher than the concentration of product in the perfectly mixed bulk. If the kinetics of the system do not obey Michaelis–Menten kinetics but do obey product inhibition kinetics, internal diffusion limitation of the substrate and the product need to be taken into account in order to design and optimise the immobilised biocatalyst particle.

Acknowledgements

The Textile Technology Group at the University of Twente acknowledges the financial support of the Foundation Technology of Structured Materials in the Netherlands and of the Dutch Ministry of Economic Affairs.

4.7 References and further reading

Agrawal P.B., Nierstrasz V.A. and Warmoeskerken M.M.C.G. (2002) 'Bioscouring of cotton textiles: the structure of cotton in relation to enzymatic scouring processes,' *Proceedings of the 2nd Autex Conference*, Bruges, Belgium, 1–3 July, p. 562.

Apfel R.E. (1981) *Methods in Experimental Physics*, ed Edmonds P.D. Academic Press, New York, Vol.19 pp. 355–413.

Aris R. (1965) 'A normalisation for the Thiele modulus,' *Ind. Eng. Chem. Fundamental*, **4**, 227–229.

Bayer Textile Processing Chemicals (Product Guide).

Bird R.B., Stewart W.E. and Lightfoot E.N. (1960) *Transport Phenomena*, Wiley, New York.

Bischoff K.B. (1965) 'Effectiveness factors for general rate equation forms,' *AIChE J.*, **11** (2), 351–355.

Buchert J., Pere J., Puolakka A. and Nousiainen P. (2000) 'Scouring cotton with pectinases, proteases and lipases', *Textile Chem. Col. Am. Dyestuff Rep.*, **32** (5), 48–52.

Buschle-Diller G., El Mogahzy Y., Inglesby M.K. and Zeronian S.H. (1998) 'Effects of scouring with enzymes, organic solvents, and caustic soda on the properties of hydrogen peroxide bleached cotton yarn,' *Textile Res. J.*, **68** (12), 920–929.

Carslaw H.S. and Jaeger J.C. (1989) *Conduction of Heat in Solids*, Oxford University Press, Oxford.

Cavaco-Paulo A. (1998), *Processing textile fibres with enzymes: An overview*, ACS Symposium Series, **687**, pp. 180–189.

Cavaco-Paulo A., Almeida L. and Bishop D. (1996) 'Effects of agitation and endoglucanase pretreatment on the hydrolysis of cotton fabrics by a total cellulase,' *Textile Res. J.*, **66** (5), 287–294.

Cavaco-Paulo A., Cortez J. and Almeida L. (1997) 'The effect of cellulase treatment in textile washing processes,' *J. Soc. Dyers Soc.*, **117**, 17–21.

Cavaco-Paulo A., Almeida L. and Bishop D. (1998) 'Hydrolysis of cotton cellulose by engineered cellulases from *Trichoderma reesei*,' *Textile Res. J.*, **68** (4), 273–280.

Contreras R.R.D. (2001) 'Biopreparación: la nueva y avanzada forma de preparar hilo y tela (primera parte)', *Rev. Latinoamericana de Tech. Textile*, **14**, 55–60.

Cortez J.M., Ellis J. and Bishop D.P. (2001) 'Cellulase finishing of woven, cotton fabrics in jet and winch machines,' *J. Biotechnol.*, **89** (2–3), 239–245.

Cortez J.M., Ellis J. and Bishop D.P. (2002) 'Using cellulases to improve the dimensional stability of cellulose fabrics', *Textile Res. J.*, **72** (8), 673–680.

Crank J. (1956) *The Mathematics of Diffusion*, Clarendon Press, Oxford.

Csiszár E., Losonczi A., Szakács G., Rusznák I., Bezúr L. and Reicher J. (2001) 'Enzymes and chelating agents in cotton pre-treatment', *J. Biotechnol.*, **89** (2–3), 271–279.

Daniels F. and Alberty R.A. (1975) *Physical Chemistry*, 4th Edition, Wiley & Sons, New York.

Etters J.N. (1980) 'Diffusion equations made easy', *Textile Chem. Color.*, **12** (6), 140–145.

Etters J.N. (1999) 'Cotton preparation with alkaline pectinase: an environmental advance,' *Textile Chem. Color. Am. Dyestuff Rep.*, **1** (3), 33–36.

Farber P. and Dahmen N. (1998) 'About fluid flow and mass transfer in textile finishing machines', *Proceedings of FEDSM'98*, 1988 ASME Fluid Engineering Division Summer Meeting, Washington DC, June 21–25, 1–7.

Farber P., Dahmen N. and Mohaupt H. (1999) 'Virtual prototyping in textile finishing: academic pastime or industrial reality?' *Melliand Internat.*, **5** (1), 75–78.

Genencor International (Product Guide).

Gooijer H. (1998) *Flow resistance of textile materials*, PhD Thesis, University of Twente, The Netherlands.

Hartzell M.M. and Hsieh Y.L. (1998) 'Enzymatic scouring to improve cotton fabric wettability', *Textile Res. J.*, **68** (4), 233–241.

Heine E. and Höcker H. (1995) 'Enzyme treatments for wool and cotton', *Rev. Prog. Coloration*, **25**, 57–63.

Lange N.K. (2000) 'Biopreparation in action,' *Int. Dyer*, **185** (2), 18–21.

Lange N.K. and Henderson L. (2000) 'Biopreparation of cotton,' *1st Symposium on Textile Biotechnology in Textile Industry*, Povoa de Varzim, Portugal, 3–7 May.

Lenting H.B.M. and Warmoeskerken M.M.C.G. (2001a) 'Mechanism of interaction between cellulase action and applied shear force', a hypothesis, *J. Biotechnol.*, **89** (2–3), 217–226.

Lenting H.B.M. and Warmoeskerken M.M.C.G. (2001b) 'Guidelines to come to minimised tensile strength loss upon cellulase application', *J. Biotechnol.*, **89** (2–3), 227–232.

Lenting H.B.M., Zwier E. and Nierstrasz V.A. (2002) 'Identifying important parameters for a continuous bioscouring process', *Textile Res. J.*, **72** (9), 825–831.

Li Y. and Hardin I.R. (1998) 'Enzymatic scouring of cotton-surfactants, agitation and selection of enzymes', *Textile Chem. Color.*, **30** (9), 23–29.

Mason T.J. (1990) *Sonochemistry the Uses of Ultrasound in Chemistry*, Royal Society of Chemistry, Cambridge.

Mason T.J. and Lorimer J.P. (1989) *Sonochemistry, Theory, Applications and Uses of Ultrasound in Chemistry*, Ellis Horwood, Chichester.

McCall R.E., Lee E.R., Mock G.N. and Grady P.L. (1998) 'Improving dye yields of vats on cotton fabric using ultrasound', *AATCC International Conference and Exhibition*, Book of abstracts, pp. 188–194.

Mielgo I., Moreira M.T., Feijoo G. and Lema J.M. (2001) 'A packed-bed fungal bioreactor for the continuous decolourisation of azo-dyes (Orange II)', *J. Biotechnol.*, **89** (2–3), 99–106.

Mielgo I., Moreira M.T., Feijoo G. and Lema J.M. (2002) Biodegradation of a polymeric dye in a pulsed bed bioreactor by immobilised *Phanerochaete chrysosporium*, *Water Res.*, **36** (7), 1896–1901.

Moholkar V.S. (2002) *Intensification of textile treatments: sonoprocess engineering*, PhD Thesis, University of Twente, The Netherlands.

Moholkar V.S. and Pandit A.B. (2001) 'Modeling of hydrodynamic cavitation reactors: a unified approach,' *Chem. Eng. Sci.*, **56** (21–22), 6295–6302.

Moholkar V.S. and Warmoeskerken M.M.C.G. (2000) 'Mechanistic studies in ultrasonic textile washing', *AATCC Annual Book of Papers*, Section 18, pp. 1–8.

Moholkar V.S. and Warmoeskerken M.M.C.G. (2001) 'Intensification of mass transfer in textile materials', *Proceedings of the 1st Autex Conference (Technitex)*, Povoa do Varzim (Portugal), June 26–29, pp. 204–213.

Moholkar V.S. and Warmoeskerken M.M.C.G. (2002) 'Mechanistic aspects and optimization of ultrasonic washing', *AATCC Rev.*, **2** (2), 34–37.

Moholkar V.S., Rekveld S. and Warmoeskerken M.M.C.G. (2000) Modeling of the acoustic pressure fields and the distribution of the cavitation phenomena in a dual frequency sonic processor, *Ultrasonics*, **38**, 666–670.

Moholkar V.S., Huitema M., Rekveld S. and Warmoeskerken M.M.C.G. (2002) Characterization of an ultrasonic system using wavelet transforms, *Chem. Eng. Sci.*, **57** (4), 617–629.

Neppiras E.A. (1980) 'Acoustic cavitation', *Phys. Rep.*, **61**, 159–251.

Novozymes (Product Information sheet).

Oxspring D.A., McMullan G., Smyth W.F. and Marchant R. (1996) 'Decolourisation and metabolism of the reactive dye', Remazol Black B, by an immobilised microbial consortium, *Biotechnol. Lett.*, **18** (5), 527–530.

Pere J., Puolakka A., Nousiainen P. and Buchert. J. (2001) 'Action of purified *Trichoderma reesei* cellulases on cotton fibers and yarn,' *J. Biotechnol.*, **89** (2–3), 247–255.

Rathi H., Mock G.N., McCall R.E. and Grady P.L. (1997) 'Ultrasound aided open width washing of mercerised 100% cotton twill fabric,' *AATCC International Conference and Exhibition*, Book of abstracts, pp. 254–262.

Rehm H.J., Pühler A. and Stadtler P. (1996), *Biotechnology. A Multi-Volume Comprehensive Treatise. Vol. 6: Products of Primary Metabolism*, Weinheim: VCH-Verlagsgesellschaft.

Shuler M.L. and Kargi F. (2002) *Bioprocess Engineering: Basic Concepts*. 2nd Edition, Prentice Hall PTR, New York.

Stanescu M.D. (2002), 'Enzymes in textile finishing,' *Proceedings of the 2nd Autex Conference*, Bruges, Belgium, 1–3 July, 446–453.

Sun S.F. (1994) *Physical Chemistry of Macromolecules: Basic Principles and Issues*, Wiley & Sons, New York.

Suslick K.S. (1988) *Ultrasound: Its chemical, physical and biological effects*, VCH, New York.

Thakore K.A. (1990) Physico-chemical study on applying ultrasonics in textile dyeing, *Am. Dyestuff Rep.*, **79** (5), 45–47.

Thiry M.C. (2001), 'Enzymes in the toolbox,' *AATCC Rev*, **1** (8), 14–19.

Tzanov T., Calafell M., Gübitz G.M. and Cavaco-Paulo A. (2001) 'Bio-preparation of cotton fabrics', *Enzyme Microbiol. Technol.*, **29**, 357–362.

Van den Brekel L.D.M. (1987) *Hydrodynamics and mass transfer in domestic drum type washing machines*, PhD Thesis, TU Delft, The Netherlands.

Van den Brekel L.D.M. and de Jong E.J. (1988) 'Hydrodynamics in packed textile beds', *Textile Res. J.*, **59**, 433.

Van der Donck J.C.J., So A. and Frens G. (1998) 'The influence of stretching on salt release from porous yarns', *Tenside Surfactants Detergents*, **35**, 119–122.

Van Holde K.E. (1971) *Physical Biochemistry*, Prentice Hall, New Jersey.

Van 't Riet K. and Tramper J. (1991) *Basic Bioreactor Design*, MDI, New York.

Waddell R.B. (2002) 'Bioscouring of cotton: commercial applications of alkaline stable pectinase', *AATCC Rev.*, **2** (4), 28–30.

Warmoeskerken M.M.C.G., Haverkamp J., Simon W. and Van der Vlist P. (1994) *A process for cleaning articles whereby soiled articles are immersed in an enzymatic aqueous cleaning medium and radiated with ultrasonic energy, characterized in that the enzymatic aqueous cleaning medium contains an enzyme having lipolytic activity*, Patent, WO9407989.

Warmoeskerken M.M.C.G., Van der Vlist P., Moholkar V.S. and Nierstrasz V.A. (2002), 'Laundry process intensification by ultrasound', *Colloids and Surfaces A*: Physicochem. Eng. Aspects, **210** (2–3), 277–285.

Warmoeskerken M.M.C.G. and Boom R.M. (1999) *90th AOCS Annual Meeting & Expo*, May 9–12, Orlando, USA.

Yachmenev V.G., Blanchard E.J. and Lambert A.H. (1998) 'Study of the influence of ultrasound on enzymatic treatment of cotton fabric', *AATCC Book of Papers*, 472–481.

Yachmenev V.G., Blanchard E.J. and Lambert A.H. (1999) Study of the influence of ultrasound on enzymatic treatment of cotton fabric, *Textile Chem. Color. Am. Dyestuff Rep.*, **1** (1), 47–51.

Yachmenev V.G., Bertoniere N.R. and Blanchard J. (2001) 'Effect of sonification on cotton preparation with alkaline pectinase', *Textile Res. J.*, **71** (6), 527–533.

5
Practical aspects of handling enzymes

H. B. M. LENTING
TNO Institute for Industrial Technology, The Netherlands

5.1 Introduction

Processes for natural fibre-based fabrics have been developed using enzyme technology for the degradation of starch after weaving, the scouring of cotton fabric before dyeing, the removal of excess hydrogen peroxide before dyeing, modification of cotton fabric (finishing or biopolishing), production of Lyocell fabric, ageing of denim, modification of wool, degumming of silk and for the treatment of water effluent from textile production mills (refer to Chapters 3 and 6 for more extended descriptions). Today, research focus is even directed to the modification of synthetic fibres such as polyester and nylon.

The application of enzyme technology is very specifically targeted to a component (substrate) present in the fibre. In this way, the main characteristics of a fibre are maintained, instead of what is experienced often with chemical processes where the fibres are modified rather unspecifically. For instance, in the enzymatic scouring process with pectinase the enzymatic action is specifically targeted towards the pectin polymer, leaving the cellulose polymers unmodified.

Besides its effectiveness, enzyme technology is also preferred for its environmentally friendly character since no hazardous chemicals are used, unlike the situation in chemical processes, and the enzyme itself is also fully biodegradable. Owing to the ongoing integration of enzyme technology in an already large, and still increasing, number of partial processes, the textile production industry is shifting from a notoriously polluting industry to a more environmentally friendly one. Additionally, the textile area has been recognised as innovative.

Many people in the textile area and working with enzyme technology are textile engineers, having no, or only limited, biochemical background. Experience of handling a biocatalyst correctly is often not available in laboratories and at production sites. The incorporation of this more practical chapter is meant to compensate for this lack in knowledge concerning enzyme handling.

The main focus of this chapter concerns the enzyme's structure stability, the different possible inactivation processes and the methods available to stabilise the enzyme. Additionally, attention is given to handling enzymes in practical circumstances and to safety issues when working with this technology.

5.2 Enzyme activity

Enzymes are proteins specialised to catalyse (biological) reactions. They are among the most remarkable bioactive molecules known because of their extraordinary specificity (concerning both substrate and reaction) and catalytic power, which are superior to those of many synthetic catalysts. In general, a chemical reaction takes place when a reactant can pass the transition state wherein the reactant possesses enough energy to form or to break chemical bonds. The difference in energy between the initial and transition state of the reactant is lowered when the reaction is catalysed by an enzyme, owing to a transient combination with the reactant. In this way, an enzyme accelerates a chemical reaction by lowering the energy of activation (see also Chapter 3).

The chemical reaction takes place in the active site of the enzyme, a three-dimensional hole in the protein where the reactant can interact optimally with one of the amino acid side chains. The three-dimensional structure of the active site should be maintained for optimal catalytic power and substrate specificity. Or, to put it another way, conservation of the enzyme specificity and catalytic power, its stability, is realised by maintaining the three-dimensional structure of the enzyme itself and its active site in particular. The overall three-dimensional protein structure is dependent on the amino acids present in the protein and the sequence of those residues. Different substructures are distinguished: primary (sequence of amino acid residues in the polypeptide chain), secondary (regular recurring arrangement in space of the polypeptide chain along one dimension), tertiary (polypeptide folding in three dimensions) and quaternary (arrangement of two or more polypeptide chains in relation to the others to form one protein).

5.2.1 Maintaining the three-dimensional structure of a protein

Different interactions between the side chains of different amino acids are the basis of the conformation and stability of the enzyme. The following interactions can be distinguished:

- Hydrophobic interaction. Some amino acids have aliphatic (alanine, leucine, isoleucine, valine and proline) or aromatic (phenylalanine and tryptophan) side chains, which can all be classified as hydrophobic. Those amino acids are often on the inside of the protein, away from the surface which is in contact with water. Those side chains are interacting under release of bound water with a subsequent increase in entropy (Edelstein, 1973). So, hydrophobic interaction is an entropy-driven process.

- Van der Waals interaction. Van der Waals interactions are caused by attractive forces between transient dipoles of uncharged atoms and are very weak. The forces operate at relatively long ranges.
- Hydrogen bonds. Hydrogen bonds (amide–carboxyl bonds in the basic polypeptide chain) contribute significantly to the stabilisation of the secondary structures of a protein, i.e. α-helici and β-sheets. Additionally, other hydrogen bonds are also important for the stability of a protein. Amino acids that can act as donors of hydrogen atoms for the formation of hydrogen bonds are the hydroxyl group-containing serine, threonine and tyrosine, and the amino group-containing asparagine and glutamine. Asparagine and glutamine can also act as hydrogen acceptors. Many of these residues in various combinations occur in the diversity of hydrogen bonds throughout the protein structure.
- Electrostatic interactions. Electrostatic ion-pair interactions often play an important role in relation to the stability of a protein. Amino acids involved are the carboxyl side chain-containing aspartic and glutamic acids and the amino side chain-containing lysine, arginine and also histidine.
- Ion coordination interaction. As a variant of electrostatic interaction, some proteins contain a cation (examples are Fe^{2+}, Mg^{2+} and especially Ca^{2+}) which functions as a structure coordinator by interaction particularly with aspartic and glutamic acids. The presence of such a coordinating ion is essential in most cases. Upon removal by extraction with a chelator, most enzymes are inactivated immediately.
- Covalent coupling of two cysteine amino acids. Two cysteine amino acids from different places in the primary structure of the polypeptide chain can be coupled under the formation of a disulphide bond or bridge. Because this coupling is covalent, it has a major impact in relation to the maintenance of the biological active conformation of an enzyme.

The three-dimensional structure of an enzyme is not a stiff conformation, but more a dynamic equilibrium wherein each interaction, as mentioned before, plays its own specific role. An impact on the three-dimensional structure can be expected when the environmental situation of an enzyme changes. As an example, when the pH of an enzyme solution is shifted, one can expect an influence on the strength of the electrostatic interactions (the degree of ionisation will be shifted), which may have an impact on structure and therefore on the activity of the enzyme. In practice, the impact will be small, if not negligible, when such a pH shift is limited, but can be dramatically increased in a situation of extreme shift.

In a situation of increased temperature, the enzyme molecules are in a higher energetic state, resulting in increased vibrations between atomic bonds. Different structure-determining interactions (the van der Waals and hydrogen interactions) will be influenced. Raising the temperature beyond a certain threshold value, which is different for each type of enzyme, will lead to loss of interactions and, therefore, protein unfolding and loss of catalytic activity.

5.2.2 Enzyme inactivation

Since the catalytic power of an enzyme depends on a correct three-dimensional conformation of protein, the active site part in particular, the biocatalyst is rather sensitive to influences from its environment which have an impact on this conformation. Changes in conformation of the enzyme often, but not always, result in reduction of its catalytic power towards zero.

Because of the complexity of an enzyme protein, many protein conformations or foldings are possible, all exhibiting their own level of free energy. Only one conformation possesses the lowest level of free energy, the native enzyme conformation. In general, inactivation of an enzyme by protein unfolding can be described as a two-step process in which the irreversible conformation at the end is reached via a transient reversible conformational change. This reversible intermediate conformation is more labile and therefore easily transforms to a more unfolded, irreversible conformation, a situation without any catalytic power. The increased lability of the transient intermediate conformation can be best explained by the increased level of free energy, see Fig. 5.1. Owing to this increased free energy level, the required free energy of activation to transfer to an inactivated enzyme conformation is lower. The reversed transfer from the inactive conformation back to the native one is blocked by the height of the energy barrier, explaining the irreversible character of this conformational change. The energy barrier for a direct transfer from the native enzyme conformation to the irreversible unfolded enzyme conformation behaves similarly.

Enzyme inactivation by protein denaturation (unfolding), aggregation (interaction between different enzyme molecules) and precipitation (protein becomes water-insoluble upon conformational change) always proceeds via an intermediate, reversible unfolded conformation. Enzyme inactivation processes upon chemical modification (covalently) may proceed via such an intermediate conformation, but direct modification of the native protein is also possible. In the latter case, protein modification will be limited towards the amino acid residues exposed to its surface.

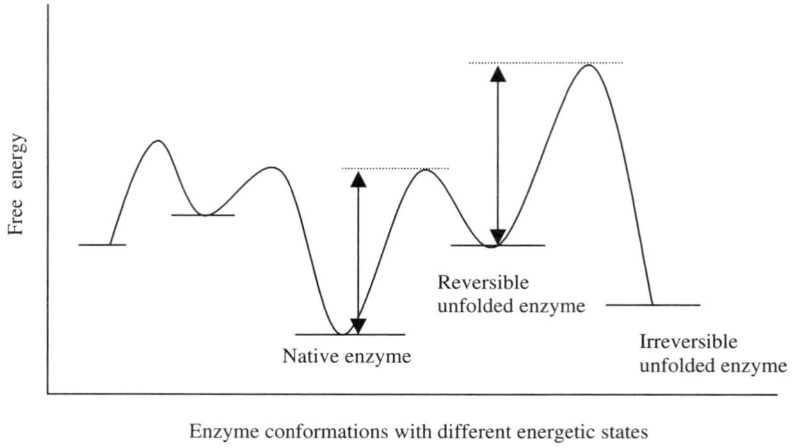

5.1 Protein unfolding in relation to the energetic state of the conformation. Arrows indicate level of activation energy required for each transition from native to irreversible unfolded enzyme.

There are many circumstances in which enzymes are irreversibly inactivated. Although this statement suggests that industrial application of enzyme technology is not realistic, the opposite is valid. A prerequisite to obtaining good enzyme technology performance is that certain circumstances have to be avoided. The following is an enumeration of what may be experienced in practice when setting different parameters and what has to be avoided.

5.2.2.1 Temperature

As in most chemical reactions, the rate of enzyme-catalysed reactions generally increases with temperature within the temperature range in which the enzyme is stable and retains full activity. The rate of most enzymatic reactions roughly doubles for each 10°C rise in temperature. The exact rate improvement varies from enzyme to enzyme and is dependent on the energy of activation of the catalysed reaction (the height of the energy barrier to the transition state). The activity versus temperature curves, as published by the enzyme manufacturers in the delivered data sheets, always show a temperature optimum. This is due to the fact that the enzyme is denaturated (unfolded) above a certain threshold temperature (each enzyme exhibits its own threshold value). The apparent temperature optimum is thus the result of two processes: the increase in reaction rate with temperature and the increasing rate of thermal denaturation of the enzyme above a critical temperature value. Mostly, the increase in denaturation rate is very high beyond the critical temperature value. The published temperature optimum is also dependent on the conditions of the test. Measurement time is an important parameter: the shorter the incubation time, the lower the possible impact of enzyme denaturation. The use of different types of substrate (reaction reactant) may also have an impact on the observed temperature optimum as a consequence of substrate binding (stabilisation factor). Therefore, determination of the optimal temperature under identical conditions to the application itself is recommended. If long incubation times are required, it is wise to fix the temperature approximately 5°C lower than the published optimum.

5.2.2.2 *pH*

Enzymes have a characteristic pH at which their activity is maximal. This maximum can be in the acidic, neutral or alkaline pH range. Often a bell-shaped activity–pH curve is observed. The exact shape however can vary considerably in form from enzyme to enzyme. The pH–activity relationship of any given enzyme depends on the acid–base behaviour of both enzyme and substrate. One can imagine that upon shift of the pH, the electrostatic interactions may change within the enzyme with subsequent possible impact on its three-dimensional conformation and thus its activity. Likewise this is true when the substrate binding to the enzyme is driven/influenced by electrostatic interactions.

The shape of the pH–activity profile usually also varies with substrate concentration, since the affinity constant K_M of most enzymes changes with pH. Such a pH–activity curve is therefore most meaningful for industrial application if the enzyme has been kept saturated with the substrate at all the pH values tested.

Most enzymes have a certain pH tolerance range, often wider than one pH unit, in which their activity is 80–100% of its maximum. Within this range, the pH may shift without major consequence for the enzymatic activity. Extreme pH values far from the enzyme's maximum in the pH–activity profile should be avoided at all times. These values have a major impact on the electrostatic interactions within the enzyme, leading to irreversible protein unfolding and thus inactivation.

5.2.2.3 *Surfactant*

When dealing with water-insoluble substrates, as is often the case in the textile production industry, surfactants are used to improve the enzyme–substrate interaction. There are four major classes of surfactants:

anionic, cationic, zwitterionic and non-ionic, all containing a more or less apolar tail or piece in the middle. All the charged surfactants can interact with the charged side chains of amino acids on the protein surface. Upon electrostatic interaction between surfactant and enzyme, a process may be initiated wherein the apolar part of the surfactant penetrates to the interior of the protein in order to accomplish hydrophobic interaction. The consequence of such interactions may be a reversible conformational change in the first instance, followed by an irreversible one. In this way, the presence of a surfactant, especially the anionic and cationic ones, results in inactivation of the enzyme. The compatibility of any surfactant has therefore to be checked prior to use. In general, the compatibility with enzymes of non-ionic surfactants is highest, whereas that for anionic is lowest. It has to be said that if one representative of a category is not compatible with an enzyme, that does not mean that another representative of that category is also incompatible. The type of apolar tail and especially the length of the apolar tail are also of importance if the above-mentioned inactivation process occurs. If an enzymatic process is developed wherein incorporation of a surfactant is required for optimal performance, a compatibility study is recommended. If no compatible surfactant can be identified, a compatible enzyme of different origin may be used.

5.2.2.4 Chelator

As mentioned in Section 5.2.1, some enzymes contain a cation to coordinate the three-dimensional structure of the protein or to act as an essential cofactor for the catalytic reaction. When such an enzyme is used, the application of strong chelators has to be avoided in order to prevent extraction of this essential ion. When a chelator is desired for a certain process, one can look for a milder variant, which still fulfills the job. As an alternative, one can look for an enzyme without such a coordinating ion or an enzyme, which binds the cation more strongly and therefore cannot be extracted by the desired chelator any more. Information concerning the presence of such a cation can be obtained from the manufacturer.

5.2.2.5 Reducing agent

When operating a process with an enzyme containing disulphide bond(s), the presence of a reducing agent should be avoided. Normally, no reducing agents are used in the processes applied in the textile production industry. Nevertheless, when a reducing agent is present, the SH groups of the cysteine residues are regenerated, breaking the disulphide bond (Fig. 5.2). Upon breakage of this disulphide bond, the three-dimensional structure is no longer fixed by this covalent bonding. Unfolding of the protein may

$$\underset{\text{enzyme}}{X{-}S{-\!-\!-}S{-}Y} + \underset{\text{mercaptoethanol}}{2\,HOCH_2CH_2\text{-}SH} \longrightarrow \underset{\text{enzyme}}{X{-}SH \quad HS{-}Y} + \text{dithiol}$$

5.2 Breakage of the disulphide bond in an enzyme via reduction by a reducing agent, i.e. mercaptoethanol.

occur, leading to inactivation of the enzyme. In such a situation an alternative enzyme without disulphide bonding has to be used.

5.2.2.6 Oxidising agent

Several amino acid residues are sensitive towards oxidation in the presence of a bleaching agent. First of all there are the sulphur-containing cysteine and methionine residues. Second, the aromatic residues tryptophan, tyrosine and histidine are also sensitive to oxidation, although to a much lower level. In a protein, those residues are oxidised when they are accessible to the bleaching agent: on the surface or in the active site. Oxidation of one of these residues in the active site may prevent adequate substrate binding and reduces the enzymatic activity towards zero, although the active site machinery and the three-dimensional structure of the enzyme is still intact. Oxidation of a surface amino acid residue may lead to conformational changes, with enzyme inactivation as the end result. The sensitive residues located further into the interior of the protein are protected from oxidation unless a reversible conformational change makes them accessible to the bleaching compound.

When an enzyme appears to be sensitive towards oxidation in the application, genetic engineering can be used as the preferred tool to replace such residue for a non-oxidation sensitive residue (Misset, 1993). In practice, the enzyme suppliers will do this. The user of enzyme technology can search for an alternative commercially available enzyme that is bleach resistant in relation to its activity.

5.2.2.7 Protease contamination

Enzymes are proteins and can therefore function as a substrate for protease enzymes. Upon (partial) degradation of the enzyme, its three-dimensional structure can collapse and inactivation will occur. In the case of protease enzyme application, autodigestion can occur wherein one protease molecule degrades another one. The addition of calcium may suppress this phenomenon to a certain extent. Contamination in the process by a protease may occur via air dust, when protease incubation has previously been executed, or by contamination with microorganisms producing extracellular proteases. In both situations, water-based cleaning of the equipment at a temperature beyond the enzyme stability threshold value is the preferred method. This can be combined with a cleaning solution exhibiting an extreme low or high pH value, conditions wherein most enzymes have drastically reduced stability. Inactivation of the proteolytic activity by chemical modification of the protease is an alternative (see Section 5.2.2.11).

5.2.2.8 Precipitation

Protein solubility in water depends on the pH of the buffer system. As mentioned before, a substantial part of amino acids with hydrophilic side chain residues are situated at the surface of the protein, for example the basic and acidic ones. The overall electric charge of the protein at a certain pH depends on the composition of those amino acids at its surface. The pH at which a protein is least soluble, and will precipitate most easily, is its iso-electric pH, defined as that pH at which the molecule has no net electric charge. At this pH, and in the range closely surrounding it (roughly a 0.5 pH unit width, although this is different for each protein), the solubility of the protein is minimal. Outside of this range, the solubility increases drastically.

In a situation where an enzyme is not fully soluble at the required application dosage, precipitation of the enzyme will occur, with consequent loss of enzymatic activity in solution. To prevent such a situation, a pH shift of several tenths of a pH unit is recommended. Depending on the pH–activity curve of the particular enzyme, the pH shift has to be realised at a lower or higher value.

An alternative practical solution of the problem of enzyme precipitation is the addition of salt, since the presence of some salt will increase the solubility of proteins. This phenomenon is called salting-in (Lehninger, 1975) and is also valid around the iso-electric point of an enzyme. A salt concentration in the millimolar range (20–50 mM is sufficient in most cases) will increase the solubility of a protein by a factor beyond 10 for most proteins. Different types of salts can be used. In general, the salts of divalent ions, such as $MgCl_2$ and $(NH_4)_2SO_4$, are far more effective than salts of monovalent ions such as NaCl, NH_4Cl and KCl. It is recommended that the impact of salt addition on enzyme performance be evaluated with respect to enzyme precipitation. Besides this salting-in principle, there is also a salting-out effect: when the ionic strength (salt concentration) is increased further, the solubility of the protein begins to decrease (see Fig. 5.3). At sufficiently high salt concentration, an enzyme protein may be almost completely precipitated from the solution. The mechanism or physicochemical basis of this phenomenon is rather complex, but one factor is that high salt concentrations may remove water of hydration from the protein, resulting in reduced solubility. In this way precipitated enzymes retain their native conformation and can be dissolved again, usually with no loss of activity. The salting-out principle may be used to recover the enzyme after applica-

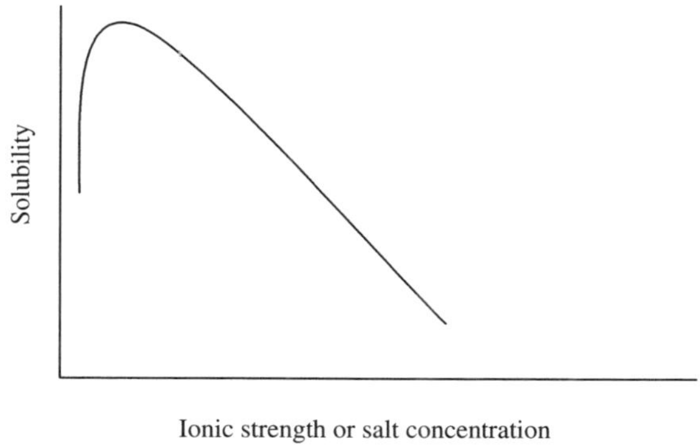

5.3 Impact of salt concentration (ionic strength) on protein solubility.

tion for re-use. Ammonium sulphate is the preferred salt because of its high solubility in water and therefore the possibility of precipitating the enzyme at almost 100%. The required salt concentration is within the molar range.

5.2.2.9 Aggregation

Normally, in a protein solution all molecules have an identical net charge of the same sign. The molecules therefore repel each other preventing coalescence of single molecules into aggregates. In a situation where a protein shows hardly any net charge, i.e. near or at the iso-electric point of the protein, repellence may not be high enough and aggregates can be formed. Depending on the type of aggregates formed, the enzyme may become partly reversibly inactivated with a significant drop in performance in the application. As in the situation of enzyme precipitation, a pH shift can be very effective in preventing this situation.

Enzyme aggregation may also occur when dealing with a protein with organised positive and negative charged side chain residues on its surface, resulting in distinguishable areas with opposite overall net charge. As a result of electrostatic interaction, dimers, trimers or tetramers may be formed with a decrease of enzyme activity as a possible consequence. In such a situation, the addition of small amounts of salt can prevent the formation of these paired enzyme molecules.

5.2.2.10 Shear force

The correct three-dimensional structure of an enzyme for enzymatic activity should be defined by the total number of interactions present in such protein (see Section 5.2.1). The overall level of interactions within such enzyme determines its sensitivity towards external added energy such as temperature. Energy addition in the form of applied shear force (stirring for instance) may also have an impact on maintaining the enzymatic activity. Shear force may initiate reversible protein unfolding first, followed by subsequent irreversible conformational change, leading to denaturation and inactivation of the enzyme. Enzymes exhibiting high turnover numbers often have a relatively low resistance against shear force, as do enzymes originating from cold-loving microorganisms. Equipment used in the textile production industry, which operates in the batch mode, has a relatively high level of intrinsic shear force compared to continuous operating equipment. In a situation where the enzyme is inactivated as a consequence of applied shear force, one can try to lower the level of shear force or to exchange the enzyme for another one of the same type but originating from a different microorganism.

5.2.2.11 Chemical modification

Chemical modification of enzymes leads to covalent changes in or on the surface of the protein, often with a shift in activity or enzymatic part-inactivation (formation of a cripple enzyme) as a consequence. Examples of covalent enzyme modification have already been mentioned: oxidation, reduction and proteolysis. Other examples are deamidation, Maillard reactions and modification of an active site residue resulting in the blockade of the biocatalyst. As an example of an active site modification, the phosphorylation of an active site serine should be mentioned (Lehninger, 1975): the hydroxyl group of an active site serine, as present in, among others, trypsin, chymotrypsin and several esterases, becomes phosphorylated upon reaction with di-isopropylphosphofluoridate. This chemical modification reduces the enzymatic activity towards zero, although the three-dimensional structure is still intact. However, the substrate is not subsequently able to bind in a correct way in the active site in order to undergo the catalytic reaction. Situations in which enzyme modification can occur, no matter what the kind of modification, should be avoided at all times. There is no simple handling possible to overcome this modification except to avoid the situation wherein such a modification agent is present.

In one particular circumstance beneficial use can be made of the active site modification. In a situation of protease contamination, the proteolytic activity may be reduced specifically by modification of its active site. A prerequisite is that the applicable enzyme should have a catalytic site different from that of the protease.

5.2.2.12 General remarks

During application, inactivation of the biocatalyst is undesirable. Subsequently, however, the enzyme frequently has to be inactivated to prevent undesirable prolonged enzymatic activity, such as in the case of ongoing cellulose material destruction by cellulase enzymes. The easiest method of effective enzyme inactivation, valid for most commercially available enzymes, is a combined treatment of extreme pH value at high temperature (80 to 90°C) for a limited amount of time (5 to 10 minutes).

A second point is that enzyme inactivation is not identical to reversible enzyme inhibition. Although in both cases no enzymatic activity is observed, the underlying basis differs fundamentally. In the situation of reversible enzyme inhibition, the inhibitor competes with the substrate to bind in the active site of the enzyme. Upon inhibitor binding, the inhibitor remains unchanged, while the enzyme–inhibitor complex prevents substrate modification (enzyme inactivation). However, the enzyme remains active, in principle; upon removal of the inhibitor, for example by dilution, the enzyme returns to full activity. Inhibition deals with competition between a substrate and a non-enzymatic sensitive substrate analogue, while inactivation deals with modification of the enzyme itself.

5.3 Stabilisation of enzymatic activity

Enzymes can be isolated for use from many natural sources. However, for industrial application, the required enzyme quantities are such that they have to be produced artificially. This is done by fermentation using microorganisms containing the gene(s) encoding the desired enzyme. The industrial production of an enzyme can be segmented into three major processing stages:

1. synthesis of the enzyme by fermentation
2. downstream processing of the fermentation broth
3. formulation of the crude enzyme product.

In the fermentation phase, the microorganisms are multiplied and induced for production and excretion of the enzyme product into the fermentation medium. In the subsequent downstreaming process the enzyme product is isolated from the broth and partially purified. Often common laboratory protein purification methods are used which have been scaled up to industrial level. Techniques used include precipitation, (ultra)filtration, various chromatography techniques and (spray) drying. The crude enzyme product derived from this downstreaming process can be either a liquid or a solid preparation. After isolation, the crude enzyme product is turned into the final enzyme product via formulation. Formulation is executed for three reasons (i) to obtain a dosage form which allows (relative) easy handling during application (ii) to show optimal performance in the application and (iii) to ensure maximal enzyme stability. The final enzyme product is stored at the production facility before distribution.

Table 5.1 Indication of typical enzyme inactivation at each life cycle stage and the duration of each stage

Life cycle stage	Duration	Typical inactivation (%)
Fermentation	Hours–days	5 to 15
Downstream processing	Hours–days	10 to 25
Formulation	Minutes–hours	3 to 10
Storage and distribution	Weeks–months	5 to 20
Application	Minutes–days	10 to 50

At the customer's site the product will be stored again, before use in the application.

In all stages of the enzyme's life cycle, the enzyme should be stable. In practice, 100% stability will not be obtained at certain stages, but inactivation of the enzyme product should be minimised as much as possible by use of all means available. In Table 5.1 typical inactivation percentages are given for each life cycle stage together with an indication of the duration of each stage. Actual inactivation percentages will be different for each enzyme, but the data in Table 5.1 give a good representative impression of how stable commercial industrial enzymes are. Inactivation already occurs in the fermentation stage as well as the accumulation or production of enzyme, as a consequence, among other things, of temperature (approximately 30–40°C) and proteolysis. The largest percentages of inactivation are seen in the application stage, especially when an enzyme will be used once in a fixed time frame and in the downstreaming stage where conditions harmful to the enzyme can be applied regularly. The relatively low percentage of enzyme inactivation during storage and distribution is striking, although the time spanned in this stage is the greatest of all.

Enzyme stability has a crucial impact on the economics of both enzyme production (supplier) and application (customer). It is therefore not surprising that major attention has been focussed on this aspect, and in particular on the formulation of the enzyme. For the customer, only two stabilities are of importance, the storage and operational stability. The storage stability and the safety issues (aerosol formation, see Section 5.5) around enzyme products are both of interest to the supplier as well as to the customer. Formulation of the crude enzyme product is applied to maximise both stability and safe handling.

5.3.1 Storage stability

An enzyme product can be delivered as a solid or liquid. In turn, a solid enzyme product can be purchased as a powder or in granulated form. The

Table 5.2 Parameters which have major impact on the storage stability of enzyme, in both solid and liquid form

Solid	Liquid
Humidity/water content	Water activity
Temperature	Temperature
	Presence of harmful additives

former physical form is less favoured, since the chance of dust formation (safety issue for employees, see Section 5.5) is substantially higher than when the enzyme is granulated.

The main parameters of importance for maximal shelf (storage) stability are summarised in Table 5.2 for both a liquid and a solid enzyme preparation. For the enzyme preparation in liquid form, the water activity is of major importance with respect to microbial hygiene. Since enzymes are natural proteins, they are biodegradable and readily used as nutrients for microbial growth. This microbial growth is dependent on water and if the conditions are favourable, the enzymes will be degraded and enzymatic activity destroyed by the microbes. The process will have an intrinsic acceleration rate since the growth of microbes will increase upon increasing the amount of digestible nutrients (partly degraded enzymes).

Water activity is of less importance for a solid enzyme product. However, a solid enzyme also contains some free water. Also the enzyme molecules contain a minimal amount of bound water (hydration), essential to maintain the active three-dimensional structure. If the solid enzyme product is stored in an environment with high humidity, the product will take up water to increase the level of hydration of the enzyme and the filler that exists in a formulated solid enzyme product. In a situation where a sufficiently large amount of water is bound to the enzyme, the enzyme may undergo structural rearrangement depending on the storage conditions, with inactivation of the enzyme as a consequence.

Temperature is a crucial storage parameter for both the solid and liquid enzyme product forms. By raising the temperature, the intrinsic energy level of enzyme molecules is also increased, with an increased level of vibration of atomic bonds within the molecules as a consequence. In the presence of a minimal amount of water (often a prerequisite) the rate of structural rearrangement is increased and thus the rate of inactivation.

Finally, for the liquid product form, the presence of potentially harmful additives is of major importance. If enough water is present, chemical reactions and interactions with, for example, a surfactant may occur that may have impact on the enzyme activity.

There are many principles or methods that may be used to achieve maximum enzyme stability either by strengthening the intrinsic stability or by minimising enzyme inactivation. The following is an enumeration of the most frequently used methods to achieve optimal storage stability.

5.3.1.1 Temperature

To start with the simplest way of decreasing the rate of inactivation, the storage temperature should be as low as possible in order to slow down all possible processes which lead to enzyme inactivation. For liquid-formulated enzyme products in particular, where enough water is present, this is an important tool to control the rate of inactivation. For this reason the enzyme supplier often recommends a maximum storage temperature. For solid enzyme products this is mostly 25°C (77°F) or below. For liquid product forms the recommended storage temperature depends on the enzyme in question. For some relatively unstable enzymes, the recommended storage temperature is below 4°C, while for others the same temperature as that valid for solid products is recommended. Therefore, the supplier's product datasheet should be read carefully directly upon delivery of the product. A general recommendation is that the storage temperature of enzymes in a liquid form should never be below 0°C, the freezing point of water. Upon crystallisation of water, the three-dimensional structure of the enzyme may be disrupted with consequences for its activity.

Exceeding the recommended storage temperature does not mean that the enzyme is fully inactivated directly. A short storage time at a temperature beyond the recommended one will raise the potential inactivation rate for the time being, but on an absolute level, the enzyme inactivation will still be limited and the enzyme product will still be valid for application. Nevertheless, storage for prolonged times beyond the recommended temperature should be avoided in order to prevent substantial loss in activity of the enzyme product.

5.3.1.2 Storage in closed containers

Another relatively simple way of preventing enzyme inactivation upon storage is to keep the enzyme in closed (plastic) containers. In this way, no water uptake can take place to initiate enzyme inactivation. Additionally, especially for the liquid enzyme products, no contamination with microorganisms that are producing and excreting proteases can occur with subsequent proteolysis of the enzyme.

5.3.1.3 pH

As is well known, every enzyme has an optimal pH for activity, as well as for stability. To store enzymes in the liquid form, the pH should be adjusted to the optimal stability region which guarantees a long shelf life. The pH can be set by the simple addition of a suitable acid, for example acetic acid or a weak buffer system. The buffer system should be weak because a pH shift should be realisable when, in the application, another pH level is required. If a choice is possible in relation to shelf life maximisation, a pH shift towards the acidic region is preferred since potential microbial growth is minimised. For maximal prevention of microbial growth, the pH should be as far in the acidic region as the chemical stability of the enzyme allows, because most bacteria will require more neutral conditions for growth. On the other hand, enzyme precipitation may begin when the pH is lowered too far. This situation has to be avoided even if the enzyme could tolerate it; in such a situation, a homogeneous enzyme dosage in the application is hard to realise, with potential consequences for the overall performance.

5.3.1.4 Reduction in water activity

As already mentioned the water activity of liquid enzyme products is of importance in relation to microbial growth in case of contamination and should be kept at a level which does not allow vigorous growth. Reduction in water activity can be realised by maintaining a high dry substance level and high osmotic pressure of the enzyme-containing liquid.

Normally, in an enzyme-containing liquid there is only 1–10% enzyme proteins. To prevent growth of contaminating microorganisms, a dry substance level of approximately 50% is required (Auterinen, 2002). Realisation of this percentage should be done, by addition of chemical additives, in such a way that the osmotic pressure of the liquid is increased as much as possible. The addition of mineral salts is the easiest way to increase the osmotic pressure. Any salt may be used for this purpose, but common salts such as NaCl are used most frequently. The amount of salt applicable has to be found experimentally for each particular enzyme since the enzyme may precipitate, either immediately or after some period of storage. There is a pH dependency of this precipitation behaviour of the enzyme. In practice, the salt concentration in a liquid enzyme preparation is between 5–10% of the weight of the final product.

A different general method of increasing the dry substance of the liquid enzyme product is to add sugars and/or sugar alcohols. The smaller the molecular sizes of the sugar, the higher the osmotic pressure it causes. Thus monosaccharides such as glucose and sorbitol are often used. In addition, sorbitol has been found to have good enzyme-stabilising characteristics. It probably chemically stabilises the enzyme molecules owing to the presence of multiple hydroxyl groups that can support the maintenance of the three-dimensional enzyme protein structure by forming ionic bonds with it. Chemicals with sugar alcohol-type structures and low molecular weight are also used, for example glycerol, ethylene glycol and propylene glycol. The amounts of sugars and sugar alcohols together that can be found in liquid enzyme formulations vary between 10 to 40% of the weight of the final product. In theory, organic solvents such as ethanol, acetone, methanol and so on may also be used to reduce water activity. However, in practice they are hardly used.

5.3.1.5 Use of enzyme interactive additives

Stabilisation of the enzymatic activity during storage by means of conformation fixation may be realised by addition of agents which bind to the enzyme molecule. One example is the addition of a substrate or substrate analogue which binds at the active site of the enzyme thus maintaining the correct and active three-dimensional structure during storage. Owing to an environment of low water activity (in the case of hydrolytic enzymes where water is one of the reactants) and the low storage temperature, catalytic reaction in the situation with a real substrate proceeds very slowly. The protective effect of a substrate (analogue) will therefore be maintained for prolonged times. Upon application, the enzyme solution is diluted, reducing the possible impact of this protective agent on the enzyme's performance. The addition of different kinds of polymers, like polyvinylalcohol, polyvinylpyrrolidone and polyethylene glycol also causes enzyme stabilisation by binding to the enzyme surface. These types of interactive additives are used for the stabilisation of both liquid and formulated solid enzyme products.

5.3.1.6 Application of absorbers and/or scavengers

Addition of agents that bind water can be beneficial in maintaining a low level of water content. In a granulated solid enzyme product, such an agent can be of particular importance when the enzyme layer is coated with this agent. Granule-penetrating water molecules from the environment will be bound by such an agent, preventing interaction with the enzyme and possible conformational rearrangement. Enzyme stabilisation by this method will be temporary. When the water-binding capacity is exceeded, the protective character of such an agent is reduced to zero. Nevertheless, depending on the concentration used, such an agent can protect the enzyme activity during a certain time window.

Using the same methodology can prevent enzyme inactivation by other harmful agents, for example bleaching agents. In the case of bleaching agents, easy oxidising molecules (scavengers) can be used to surround the enzyme layer. In this way neutralisation of the bleaching agents occurs before the enzyme is reached.

5.3.1.7 Incorporation of salts

Calcium salts or ammonium sulphate may also be used to protect enzymatic activity during storage. In liquid formulations, calcium salts are used in concentrations of less than one molar. The exact stabilisation mechanism is not well understood, but probably coordination by calcium ions (electrostatic interactions) in one way or another will strengthen the enzyme protein structure. Ammonium sulphate is often used for salting-out proteins (see Section 5.2.2.8). In the precipitated form, enzymes are quite inert towards any agent or physical condition. Upon dilution after storage, the enzyme returns in solution and will show its original activity. In principle, ammonium sulphate precipitated enzyme can be stored in both solid and 'liquid' form. There is one potential disadvantage in applying this method: upon resolvation by dilution, a relatively large quantity of salt is present, which may have an impact on the enzyme performance. In this case, removal of the salt is required prior to application, and this can be achieved by use of size-exclusion chromatography or dialysation, for example.

5.3.1.8 Addition of microbicides

To prevent loss of enzymatic activity by contamination with, and growth of, microorganisms which produce proteases, the incorporation of microbicides is another option as well as reduction of water activity. The most frequently used microbicides for enzymes are salts of benzoic acid and sorbic acid. Both microbicides are most protective in a mild acidic medium: the benzoate at around pH 4–5, the sorbate at around pH 5–6. The amount of benzoate or sorbate required varies between 0.1 and 0.3% of the weight of the final product. Alternatively, the methyl-, ethyl- and propyl esters of benzoic acid, called parabens, may also be used. These agents are optimally protective in the more neutral pH regions 6–7. The main disadvantage of this class of agent is poor solubility in water. Propyl paraben can be added only at 100 ppm (parts per million), methyl paraben at around 1500 ppm. All mentioned microbicides are compatible with enzymes and normal additives in food and thus their use is very safe.

There are more non-food applicable microbicides and some of them may be very effective for protecting enzyme preparations from contamination during storage. Their compatibility with the enzymes used in textile industrial processes has to be checked prior to incorporation in enzyme preparations.

5.3.1.9 Enzyme immobilisation

Fixation of the enzyme onto a carrier will have a stabilising effect in relation to its three-dimensional structure. In particular, multi-point attachment, fixation via several covalent bonds, often leads to a drastic increase in conformation structure stability and therefore activity stability. However, the immobilisation treatment itself normally already leads to an activity decrease of roughly 10 to 20%. If the immobilisation is on the surface as well as in the interior of the carrier, (substrate and product) mass transfer limitation may occur, depending on the carrier particle size. In a situation of mass transfer limitation, part of the enzymatic activity capacity will not be used in the application (lack of substrate availability) leading to an observed apparent inactivation of the enzyme. When there is prolonged inactivation of the enzyme at the outer areas of the carrier during application, the intact but initially unused enzyme at the more internal areas of the carrier becomes active when the substrate reaches it after passing inactivated enzyme at the outer sites. Finally, immobilisation will raise the cost price of the enzyme system. Therefore, in practice, this technique for enzyme stabilisation during storage is not used.

Nevertheless, the technique is powerful in those circumstances where the enzyme can be reused in the application many times in order to reduce enzyme costs. Examples, where enzyme immobilisation is applied successfully are in the conversion of glucose into fructose for the production of high-fructose corn syrup by use of immobilised glucose isomerase, and in the production of 6-aminopenicillinic acid by use of immobilised penicillin acylase. Potential is identified for use of immobilised enzyme systems in the textile production sector in wastewater treatment and in the removal of excess hydrogen peroxide by use of immobilised catalase.

5.3.1.10 Genetic engineering

A different way of stabilising enzyme activity is to modify the enzyme protein itself by means of genetic engineering. Despite the fact that the structure–stability relationship is not always known, when the amino acid that causes instability is identified, site-directed genetic engineering could be applied. If the amino acid is not identified, random mutagenesis is a possibility. In this case many different mutants are generated randomly and screened for improved stability. Compared with site-directed mutagenesis, random mutagenesis is quite laborious and is therefore less preferred. In a situation where it is known which amino acid residue is modified upon inactivation, this amino acid may be replaced by, in principle, any other one. In practice, substitution by different amino acids often has to be studied, since not only is the modification of that particular amino acid blocked by its replacement, but also a shift in both the enzymatic activity, and, more importantly, performance is initiated. In addition, incorporation of another amino acid may result in incorrect folding of the protein, resulting in a total loss of catalytic activity. Nevertheless, most trials of modifying an enzyme through replacement by a particular sensitive amino acid result in improved stability while maintaining the performance as it was with the original wild-type enzyme. The major benefit of this method is that once the genetic code for an enzyme has been changed, production will be like that of the wild-type enzyme, and its behaviour in the downstreaming process is often similar. So the economics of producing a mutant enzyme will be identical to that of the wild-type enzyme. Performance in an application and enzymatic activity using a particular substrate are not correlated. Therefore, performance of generated mutants has to be evaluated instead of an activity test. As an example, the successful improvement of bleach stability (against hydrogen peroxide) of a protease has been realised upon replacement of the bleach-sensitive methionine by either a serine or glutamine (Misset, 1993).

5.3.1.11 Chemical modification

A final possibility for improving stability of enzymes is chemical modification. This can be a one- or two-point (internal crosslinking) modification. As an example for a one-point modification the modification with glyoxal can be mentioned. A well-known crosslinking procedure is the reaction with glutaraldehyde where the aldehyde groups at both ends of the molecule react with the amino groups in the side chain of amino acids. Chemical modification requires reactive amino acids, which are the basic, acidic, alcoholic, aromatic and the sulphur-containing ones. This modification is not selective because the reagent will modify all surface-exposed amino acids of a given type. Although improved stabilities may be realised, there are major drawbacks to this method. First, the enzymatic activity and enzyme performance may be modified (often reduced) upon chemical modification, resulting in the need for increased enzyme dosage in the application. Second, the chemical modification step has to become an integral part of the enzyme production process and will increase the cost of the enzyme product. For these reasons and the technical limitations, chemical modification is often not attractive to apply in practice.

In the textile production industry, enzyme technology has already been in use for some years. A striking observation is that roughly 70–80% of the enzymes used in this industry are in non-formulated powder form (Auterinen, 2002). Although enzyme formulation is not essential for stability *a priori*, it is essential for dust prevention (safety issue, see Section 5.5). Handling enzyme powder instead of formulated enzyme product is more dangerous for employees because the chance of dust formation and subsequent enzyme inhalation is larger. Since the use of enzyme technology is growing in the textile area, both in quantity and number of applications, more attention should be given to safety of handling and therefore to the use of formulated products. In another industrial area with even more experience of enzyme application, i.e. the detergent or cleaning area, this aspect has been given full attention. In this sector high-tech formulations have been developed in order to incorporate enzymes in a safe-handling and enzyme-stable way into detergents, which are harmful environments for enzymes with high pH, anionic surfactants and bleaching agents. Whereas Novozymes has developed the T-granulate formulation (T stands for tough), the other major enzyme supplier, Genencor International, applies its own developed Enzoguard® technology. The T-granulate, a more or less spherical particle, contains a core wherein the enzyme is incorporated. This core consists of cellulose fibres and is encapsulated by an inert coating. The material of the core and its coating give the granulate elastic properties, which makes it resistant to mechanical crushing and the release of enzyme dust.

The Enzoguard® technology is based on a different principle. By applying fluid-bed granulation technology, a multitude of layers is applied around a small spherical prefabricated core. The first layer on top of the core, containing a sugar, is the enzyme-containing layer. Owing to its concentration within a small spherical layer around the core, instead of within the core, the enzyme will dissolve in water very fast. The second layer around the enzyme layer has a scavenger function and contains chemicals to protect the enzyme from inactivation by chemicals (bleaching agents) and water (humidity) coming from outside. The outside layer is a functional coating developed and applied for optimal storage stability of the enzymatic activity.

Both formulations, although different, show substantially improved enzyme storage stability, very well reduced levels of potential dust formation upon handling and good dissolution profiles in water. It may function as a good example in the textile area of how to formulate enzyme powders for improving safe handling and storage stability.

5.3.2 Operational stability

As well as during storage, the enzyme may also be inactivated during application. Although the time window for application is relatively small compared with that of storage, the rate of inactivation is often the opposite. As in the situation with storage stability, there are some methods that maximise enzyme stabilisation during application. First of all, in application the goal should be the best obtainable enzymatic performance at a certain fixed cost level, and not enzyme stability *per se*. In a situation where the enzyme performance can be obtained rather easily, there is or may be some space left to trigger some application parameters towards improved enzyme stability. Even in a situation where the enzyme is applied once with subsequent inactivation, an improved stability can be beneficial through a reduced level of required enzyme dosage.

Certain methods that can be used for enzyme storage stability improvement are also valid for increasing its operational stability. These methods are:

- additive dosage
- immobilisation
- genetic engineering
- chemical modification.

The last three methods, used for improvement of storage stability, will also have a positive impact on operational stability. For the first method, it should be kept in mind that additives, dosed in the enzyme formulation for storage stability improvement, will lose their stabilising effect in the application as a consequence of dilution of both enzyme and stabiliser. Besides the above mentioned methods, there are some additional ones, which are enumerated below.

5.3.2.1 Temperature

In principle, decreasing the operational application temperature will lead to improved enzyme stability. Equally, the enzymatic activity will also be lowered. If the factor time is of no importance, one may screen for an optimal temperature at which the process costs are minimised (a balance between costs of chemicals and energy required). In practice, noticeable enzyme inactivation starts after passing an enzyme-particular threshold temperature. The optimal temperature recommended by the enzyme supplier is mostly a value just below this threshold value. In many screening tests for optimal operation temperature one will end up near the recommended temperatures when the intrinsic energy content of the enzyme is the driving force for inactivation (denaturation of the protein). In those situations where the enzyme inactivation is induced, for example by interaction with a compound (among other things this can be a charged surfactant) present in the application, the screened optimal temperature may be substantially different (lower) from the recommended temperature. The opposite situation can also occur: when an enzyme is stabilised by interaction with a component (among other things this can be a polymer), the threshold value at which enzyme inactivation starts can be shifted towards higher temperature values. In general, the temperature is not a parameter that is used for operational stability improvement.

5.3.2.2 pH

As mentioned in Section 5.3.1, the pH value for maximal activity is often not the value at which the enzyme shows its maximal stability. As stated in the beginning of this section, the focus in the application should be on enzyme performance, often, but not always, coupled with activity. In an application situation where the enzyme is used once and inactivated afterwards, the pH should be chosen normally for (near) maximal enzymatic activity instead of stability. However, when an enzyme-destabilising interaction with another required ingredient in the application is present, for example interaction with a charged (anionic or cationic) surfactant, it may be valuable to choose a pH for maximal enzyme stability. The rate of interaction-induced enzymatic inactivation can by very high with huge impact on the performance as consequence. Electrostatic interaction that leads to enzyme inactivation by unfolding is pH dependent. Thus, a shift in operation pH value away from that of maximal activity may increase the overall performance as a consequence of reduced enzymatic inactivation rate.

In a situation where the enzyme is reused, the enzyme stability during application becomes more important. In such situations, the pH has to be set more in relation to maximal half-life time (the time in which the enzymatic activity has been decreased by 50% while operating under application conditions). An eventual decrease in enzymatic activity can be compensated by a higher dosage or by increasing incubation time. The exact pH setting should be done in such a way that a minimum value of the overall processing costs is realised.

5.3.2.3 Substrate concentration

When an enzyme and its substrate are brought together, they initially form an enzyme—substrate complex prior to the enzymatic reaction to generate product. During the catalysed reaction, this complex is in equilibrium with the individual substrate and enzyme molecules. When increasing the substrate concentration, this equilibrium is shifted towards the enzyme—substrate complex form. In such a situation of increased substrate concentration, more enzyme will be present in the enzyme—substrate complex form. As already mentioned in Section 5.3.1.5, the binding of a substrate molecule in the active site of an enzyme may stabilise the enzyme by preventing protein unfolding. Working in the application with a raised substrate concentration will therefore stabilise the enzyme, especially at the start of enzymatic substrate conversion where high amounts of substrate molecules are still present in the system.

This method for improving the operational stability of enzymes is of less importance for the textile industry because, in textile production processes, the substrate will often be either the fabric itself or the substrate concentration will be fixed, as in the situation where the desizing procedure makes use of α-amylase. Manipulation of the substrate concentration in these situations is not possible. It might, however, be a method for the future treatment of wastewater by enzymes.

5.3.2.4 Additives

As mentioned at the start of this section, additives may have a positive effect on enzyme operational stability as well as the effect they can have on storage stability. Addition of ions in the application may have an enzyme-stabilising effect under application conditions (diluted enzyme condition) owing to electrostatic interaction, without any negative impact on its performance. As an example in the textile industry, the optimal use of many α-amylases in the desizing process can be achieved in water with a hardness of around 15°dH instead of using demineralised water. Similarly addition of other components such as diverse polymers, e.g. polyvinylalcohol or polyvinylpyrrolidone and surfactants (especially the non-ionic ones) may have a stabilising effect that is due to interaction with the enzyme, thus preventing protein unfolding.

5.4 Handling of enzymes

In principle, an enzyme should be handled as a chemical with all the same precautions necessary for handling chemicals. In addition, enzymes are bioactive as long as their three-dimensional structure is maintained. To secure this structure in solution, some additional precautions have to be taken, such as avoiding extreme circumstances concerning pH and temperature, and incompatibility with other processing agents.

For general handling of enzymes, the handling guide received from the enzyme supplier should be followed. If a particular enzyme requires a different handling procedure, this will be notified in the enzyme datasheet or in a separate document. In this section some handling of enzymes in different situations is discussed.

5.4.1 Handling enzymes in batch and continuous equipment

In general, application of enzyme technology in the textile production industry will proceed in batch or continuous operating equipment. The application and handling of enzymes in both types of equipment will differ somewhat as a consequence of the process flow within that equipment. For both situations, the preparation of the ready-to-use enzyme solution will be the same, although the final enzyme concentration may differ. The method for preparing such solutions is described in Section 5.4.2.

5.4.1.1 Batch equipment

In batch equipment, for instance in a Jet construction, a limited amount of fabric is deposited in the equipment and treated in a process where a relatively high level of shear force is applied, inherent to the equipment used. The fabric has a high level of contact with the water medium, while the flow through the fabric (mass transfer) is on a relatively high level. The processing time in such equipment can be varied very easily as it can be set on hold for a while to check the performance level.

In this type of system, the enzyme solution will be introduced to the process when all other ingredients, together with the fabric, are already there. The required enzyme dosage, as determined in previous labscale experiments, can be applied at the start of the application and the application process is executed using parameters compatible with the enzymatic activity. The stirring device is activated directly upon enzyme dosage and heating the incubation medium to the required temperature application level is started. The temperature of the ready-to-use enzyme solution itself should preferably be between 5 and 20°C. During the process, the pH should be checked regularly in order to maintain the correct pH tolerance range. If correction is required, this can be realised using one of the methods described in the next section. If the development of performance lags behind previous experience, several possibilities are available to reach the desired performance if the origin of the lower performance is related to enzymatic activity. The simplest solutions are extending the incubation time and raising the temperature. The latter option should be chosen only in the last stage of the enzyme process, since the enzyme denaturation rate will be increased parallel to that of enzymatic activity. Additional enzyme dosing directly in the application is an alternative method.

After application, dosing an alkaline agent directly into the application medium, to raise the pH beyond 10 and raising the temperature to beyond 80°C with maximal heating capacity, can inactivate the enzyme. Since this will often take a couple of minutes, a duration time of 5 minutes at this elevated temperature will be enough to inactivate the enzyme.

Applying enzyme technology in the textile production industry is done most easily in equipment operating in the batch mode as a consequence of the easy way in which the different required process parameters can be established. There is only one parameter whose level has to be verified to maintain enzyme activity, the applicable shear force. Since the shear force can be rather high is this equipment, it may be a driving force for enzyme inactivation by protein unfolding for those enzymes exhibiting a rather labile three-dimensional structure.

5.4.1.2 Continuous operating equipment

In a process operating in a continuous mode, the situation for enzyme application differs substantially from that in the batch mode. Here, the fabric substrate and enzyme medium are not normally in continuous contact during the whole process, but for a short time only. Wetted fabric with a transport velocity of 1 m/s or beyond is squeezed prior to passing through an enzyme bath. There, the fabric has to pick-up the enzyme or adsorb a film of enzyme-containing water. During incubation, the fabric may remain in a steamer or other temperature-conditioned environment. However, the incubation time is limited by the amount of fabric in such a compartment (limited capacity for fabric hold-up). If enzyme incubation occurs in the enzyme bath itself for reasons of, for instance, failure in pick-up of the enzyme by the fabric, the incubation time must be even shorter since the possibility of holding a substantial amount of fabric in the solution-containing compartment are even more limited. In the former situation, it is recommended that the volume of the enzyme bath be limited. Dosing of additional liquid enzyme solution from a cooled enzyme storage vessel can be carried out continuously while operating the process.

Monitoring the enzymatic activity may be a problem in a continuous operating enzymatic process. Monitoring is required in order to know the velocity of enzyme pick-up by the fabric and therefore to establish the enzyme dosing velocity from the storage vessel. A constant enzyme dosage level in the application is essential for many enzymes in order to obtain a homogenous performance level. In an ideal situation, the measurement of the enzymatic activity itself is preferred. However, such measurement often requires a time window of a few minutes or more, whereas on-line information is preferred for accurate enzyme dosage. This is an important issue since enzyme dosing may become essential for obtaining a homogenous performance level when the incubation time is more and more reduced.

On-line measurement of enzyme level may be done via an indirect method, i.e. by measuring the amount of protein via spectroscopy. This can be achieved by using a flow-cell unit and measuring the absorption at a wavelength of 280 nm. The aromatic tyrosine residue absorbs at this wavelength and since most proteins contain tyrosine residues, the wavelength can be used often. Where the enzyme applied does not contain this amino acid, measurement at approximately 275 nm (absorption by tryptophan) or 260 nm (absorption by phenylalanine) provides an alternative. A prerequisite for applying this method is that the measured absorption can be correlated directly with the amount of active enzyme. This means: (i) the enzyme must be applied under conditions where it maintains its full activity (no significant inactivation may occur); (ii) no other ingredients in the application should absorb at the wavelength applied; (iii) the fabric should not release any chemical/component which absorbs at this wavelength; and (iv) the enzyme preparation used should be rather pure with respect to protein content. If this last item is not the case, enzyme monitoring by this method cannot be achieved, since the level of pick-up by the fabric of both enzyme and contaminant protein may be different. As a possible consequence, a shift in performance during process operation may be observed.

In a situation where there is no specific pick-up of enzyme by the fabric but only via adsorbing a film of enzyme-containing water, the enzyme dosage is relatively simple. Under circumstances where the enzyme maintains its full activity, it leaves the enzyme bath by adsorption of enzyme solution on the squeezed fabric. Maintaining the level of enzyme solution is enough in this situation to ensure constant enzyme dosage and thus performance.

When the enzyme incubation in a continuous process has to be executed in the enzyme bath itself, a comparable, but not identical, situation occurs to that in a batch process. In the batch situation, all fabric undergoes enzymatic action at the same time. The impact of a possible shift in enzymatic activity during incubation will be experienced by the total fabric batch. In a continuous operating process, fluctuations in the enzymatic activity will not be experienced by the total fabric batch. Therefore, the enzymatic activity has to be kept constant. This can be realised by applying conditions whereby no loss of any enzymatic activity will occur. In practice, some inactivation is frequently faced in spite of maintaining stable temperature and pH conditions. Therefore, monitoring the activity is essential at all times. In this situation, measurement of the enzymatic activity on a regular basis will be accurate enough, although on-line activity measurement is still preferred.

In any case, an enzyme bath cannot be used indefinitely. Inactivated and unfolded protein will accumulate in the bath and may influence the performance of subsequent processes (dyeing and finishing) by binding or deposition on the fabric after passing a certain threshold of inactivated protein concentration. To prevent such unwanted shift in product quality, it is recommended that each continuous process be started with a new, fresh made enzyme solution on a regular basis. How many hours/days such enzyme solution can be used depends on the enzyme itself and the set of parameters applied in the process, and cannot be predicted in advance. However, one should be aware that an unexpected shift in product quality (increasing inhomogeneity) might be triggered back to the freshness of the enzyme solution used. Inactivation of the enzyme after application can be achieved by conducting the fabric through an alkaline solution followed by passage through a steamer.

5.4.2 Practical issues

Although at present, enzyme technology has been widely embraced in the textile area, the fine detail of the technology is not always known. Those employees who are working with enzymes on a daily basis, should be given practical training in how to handle enzymes safely, both for the sake of the person involved, as well as for enzyme product quality (enzymatic activity maintenance). Additionally, factory managers should also have practical knowledge of how to manage enzymes in order to create safe working facilities and environments for employees. In this section some enzyme handling situations will be discussed.

5.4.2.1 Enzyme storage

Enzymes which are delivered to the factory site, no matter in what form, have to be stored before they are used in an application. Enzymes will often, if not always, be delivered in closed plastic bags or drums, depending on the volume and type of formulation. During storage, high values of both temperature and humidity should be avoided. The environment in which the enzymes are stored should preferably be an isolated room within the factory which is kept cool, dry and well ventilated. Ventilation will remove any possible enzyme dust or aerosols generated on (un)packing enzyme products, and will secure a hygienic and safe environment for the employee. Upon storage of liquid formulated enzyme, temperatures below zero should be avoided at all times.

5.4.2.2 Enzyme preparation for application

Handling enzymes always has to be done in well-ventilated areas, no matter where in the factory. When opening a closed bag or drum containing solid enzyme product, handling procedures which generate high quantities of dust should be avoided. In the case of formulated enzyme product (granules), crushing the granules will generate dust and should also be avoided. In a situation handling enzyme powder, the transfer of enzyme powder into a different container (or whatever) should not be achieved by vigorous pouring but in a gentle way, for instance by ladling. A protective facemask, safety glasses and gloves should be used at all times during enzyme handling. After taking away the required amount of enzyme, the bag or drum has to be closed immediately in order to minimise the amount of water uptake (protecting the remaining enzyme). In handling concentrated liquid enzyme formulations, the enzyme transfer should be done under hygienic conditions to avoid contamination by microbes. For the same reason, the number of times enzyme is taken from the same drum should be limited.

To prepare a diluted enzyme solution in water for the application, no matter if the concentrated enzyme product is in the liquid or solid form, the addition of enzyme to the water medium should be done as the last step in the preparation procedure. The order of addition will be such that first all other ingredients such as buffer, surfactants, and salts are introduced and dissolved in the medium. This medium should then be brought under the desired conditions of pH and temperature, and checked before the enzyme is dosed. If spillage occurs during this enzyme preparation, no matter if it is a concentrated or dilute enzyme solution, cleaning should be done directly before drying out can take place (prevention of dust formation).

5.4.2.3 pH maintenance during application

An enzyme is catalytically active in a certain pH range, which is different for each particular enzyme. Normally, at the start of the enzyme application, the pH is set by addition of acid, base or buffer of a particular strength. During application, the pH should be regularly monitored, preferably online. In a situation where the pH has shifted, correction may be required. pH shift may occur if an acid (most common in hydrolytic reactions) or base is produced in the catalytic reaction. Correction of the pH upon shift should not always be carried out. Each enzyme exhibits a certain pH tolerance: the enzyme is active for 80% or more of its maximal obtainable activity in a certain pH range. As long as the observed pH shift is within this range, no correction is required. To optimise the use of this tolerance range, the initial pH setting may be done not in the middle, but at one border of this range, creating a maximal possible pH range in which the pH can shift during ongoing enzymatic reaction without major impact on the enzyme activity.

If a pH correction during application is required, several methods are possible. Preferably, diluted acid or base (depending on the pH shift direction) is directly added to the application medium under vigorous stirring. In this way, the local high or low pH at the place where the acid/base is introduced into the application system will be kept as short as possible and thus prevent inactivation of the enzyme by pH-induction. Alternatively, an identical buffer solution with an increased strength can be added. The disadvantage of this method is enzyme dilution in the application. The method of tapping off a small volume of the incubation medium, and addition of acid/base to this fraction prior to re-entering the application should be avoided since the enzyme in this fraction will be inactivated (the acid/base load needed to realise the required pH shift in the whole incubation system will result in an overcompensation of the pH value in this small fraction). This method can be used only in the case of immobilised enzyme systems, where the enzyme is drained off prior to acid/base addition.

In a follow up execution of this application, the use of a buffer system with increased strength is recommended once a required pH shift is experienced.

5.4.2.4 *Enzyme inactivation at closure of the application*

Often, irreversible inactivation of the enzyme at the end of the application is beneficial because ongoing enzymatic action may lead to undesired effects like extended tensile strength loss and 'overperformance' in the case of cellulase application. Inactivation of the enzyme by disruption of its three-dimensional structure can be easily realised by shifting the pH and temperature to extreme values for a relative short time period. Using normal, not thermostable enzymes in the application, a treatment for 5 to 10 minutes at a pH level above 10 and a temperature above 80°C is effective in destroying the enzymatic activity completely. Dealing with a thermostable enzyme, the treatment time and/or temperature should be increased. Shifting the pH to an extreme acidic value is also possible in principle. However, most applicable enzymes have the highest pH stability in the mild acidic range. Therefore, the biggest pH shift is realised when moving towards an extreme alkaline value. Additionally, cotton fabric is more resistant to extreme alkaline treatment in comparison to extreme acidic treatments. Other alternatives for enzyme inactivation, such as proteolysis, precipitation, chemical modification and protein unfolding inducer or enzyme inhibitor addition, are all less attractive for reasons of time, economic costs and/or efficiency level.

5.4.2.5 *Compatibility additives*

In enzyme applications, additives are regularly used for different objectives. Compatibility of those additives with the enzymatic activity must be checked prior to application because, especially in the groups of surface active agents and chelators, there are components which are not compatible. In the group of surface active agents, the charged components often have the ability to bind to oppositely charged amino acid side chain residues upon which an irreversible protein unfolding process is started-up. If the presence of a surface active agent is required, a non-ionic component is recommended.

In the situation where charged agents in the application are essential, no matter if these are anionic or cationic, the combination of such a charged agent and a non-ionic agent is recommended. Often a reduced negative impact of the charged surface active agent is observed upon application of such a mixture. One explanation for this observation is that the hydrophobic binding of the non-ionic agents to the surface of the enzyme prevents interaction with the charged agent. As an example, it is known from the literature that cellulases are sensitive to both anionic and cationic surfactants (Ueda *et al.*, 1994) whereas they are compatible with non-ionic surfactants. These non-ionic surfactants even improve the enzyme's performance upon addition in the application. After extensive study, Kaya *et al.* (1995) came to the conclusion that non-ionic surfactants hinder the binding of the enzyme on cellulose by reduction of the binding strength. In this way, the desorption rate of the enzyme is increased leading to an improved enzyme activity and performance level.

Also, for structural reasons, additives from chelators are potentially not compatible with enzymes containing a coordination ion. Each chelator has its own affinity towards a certain ion, which influences the strength of complexation. As a consequence, when the binding affinity towards an added chelator is higher than towards the protein, the ion will be extracted from the enzyme, disrupting the structure and inactivating the enzyme. If the presence of a chelator is required in the application, it needs to have a lower affinity for the enzyme-bound ion and yet still function in the application. As an example, the application of the BioPrep 3000L enzyme can be mentioned in the bioscouring process of cotton fabric. The enzyme contains a calcium ion to maintain its three-dimensional structure and, therefore, addition of a strong chelator such as EDTA (ethylenediaminetetra-acetic acid) will destroy its activity (Novozymes, 1999) whereas a mild chelator variant like STPP (sodium tripolyphosphate) is compatible with the enzyme (Lange, 1999).

5.4.2.6 Enzyme dosage

It is frequently thought that the amount of enzyme activity (dosage) and incubation time are exchangeable in a linear way in the application and that the more catalytic breakdown that has been realised, the better the performance obtained. For some enzymes in certain applications this may be valid, more or less. However, the opposite is true in many situations dealing with (insoluble) polymers as substrates, as is the case in the textile industry. Other operational conditions may be more important or essential than the actual level of enzymatic activity, and in these situations, increasing the incubation time, or raising the enzyme dosage will not result in an improved performance level. As an example, the application of cellulase on cotton fabric to produce biopolished fabric should be mentioned. Performance will be achieved only upon passing a certain threshold value of enzymatic breakdown of the cellulose fibres. Removal of the enzyme-weakened microfibres is achieved by the application of shear force above a certain minimal level (Lenting and Warmoeskerken, 2001). Under this threshold value of shear force, no microfibre removal will be realised, no matter if the enzyme dosage and/or incubation time is increased. In a situation with a shear force level above the minimal value, a higher enzyme dosage will not lead, *a priori*, to performance in a shorter time frame. Increasing the enzyme dosage, will, besides a general increase of hydrolysis velocity, result in a different pattern of cellulose hydrolysis, but these modified kinetics do not lead, *a priori*, to accelerated performance. Incubation time and enzyme dosage have to be handled in this situation as being related, but not linearly exchangeable parameters.

When a concentrated liquid enzyme product is available for application where fabric is already present in the incubation medium, it is recommended that the enzyme solution in the incubation medium is diluted prior to addition to the fabric. This procedure can prevent local overperformance and/or inhomogeneous performance since effective mixing is often not realised at the time of enzyme addition (mixing device on hold during this handling).

In a situation where a rather concentrated enzyme solution has to be prepared from granulated enzyme product, it is recommended that the granulated enzyme is dissolved in a separate vessel, followed by a filtration procedure prior to introduction in the application. In this way introduction of possible insoluble substances originating from the granules can be avoided. Such insolubles may have some impact on the substrate, such as with inhomogeneous dyeing afterwards.

5.4.2.7 Repeated enzyme use

To make the enzyme technology more profitable, re-use of enzymes should be considered since they are relatively expensive. However, to ensure performance level during re-use, circumstances should be identical in all cycles of re-use. In a situation where processing agents other than an enzyme are used, which regularly occurs in the textile production area, one has to assure that accumulation of agents does not occur. A shift in the concentration of a processing agent, for instance a surface-active agent, may result in a shift in performance as a consequence. These agents should therefore be monitored both for underdose and overdose in the subsequent enzyme application cycles. Apart from its impact on performance, a shift in processing agent concentration may also have an effect on the stability of the enzyme and therefore on its half-life time.

On re-use of an enzyme in different subsequent cycles, one has to be alert for a different phenomenon: the potential impact of denaturated enzyme protein. During application, some enzyme will be inactivated and denaturated (protein unfolding). It may be possible that, after passing a threshold value of denatured protein, this protein becomes insoluble and is deposited on the fabric. If this phenomenon occurs, it may initiate problems later on in the dyeing and finishing processes. However, in the presence of a charged surface-active agent, the chance of such a redeposition process taking place is greatly reduced. If such an agent is not already present as a processing agent, it may be applied for this reason, if compatible with the enzyme.

To re-use an enzyme and avoid the above-mentioned potential problems, the isolation of the enzyme from the medium after each application cycle should be considered. This can be done in different ways in principle, but to be profitable it should be achieved in the simplest way possible, such as size-exclusion chromatography. Enzyme immobilised on a carrier can be isolated by simple drainage of the medium over a gauze filter.

5.4.2.8 Enzymatic activity

The supplier often guarantees the amount of active enzyme in an enzyme preparation by its activity. This enzyme activity is measured on a defined substrate using a fixed set of parameters. In many cases, the substrates used are easily soluble, modified substrates; for instance carboxymethyl- or hydroxyethylcellulose when measuring cellulase activity. Synthetic chromogenic substrates (like nitrophenyl cellobiosides for cellulases) are also used for convenience and reproducibility, and the method to be used is mostly available on request. Furthermore, different enzyme suppliers use different assay systems to measure the activity of the same enzyme type; for instance that of the cellulase enzyme. Enzymatic activity is expressed in Units (the amount of micromoles of product formed in one minute) per milligram of protein or product (volume). If a preparation contains pure enzyme, the activity measured per milligram of protein is called the specific activity. Each enzyme originating from a different (microorganism) source exhibits its own specific activity, but activities measured under different circumstances and using different substrates, not being the substrate in the application itself, cannot be directly compared with each other and cannot be linked directly to a certain performance level. Therefore, replacement of an enzyme by another of the same type but different origin in an application, maintaining all the parameters including enzyme dosage, as expressed in Units, does not, *a priori*, result in the same performance.

A given enzymatic activity for an enzyme preparation should be used only for quantification of the amount of active enzyme. For determination of the required dosage in the application, an activity–performance profile should be generated under the conditions of the real application. Likewise, a pH-activity profile, as given in the data sheet of an enzyme product, cannot be read as a pH–performance curve; such curve should be measured under the conditions of the application itself while monitoring performance instead of activity. The same is valid for the temperature–activity profile, especially when dealing with insoluble substrates. Upon binding of the enzyme to its substrate, the temperature stability may be improved. Often in the textile area, the substrates used are insoluble.

The last point for consideration is the fact that each different batch of the same enzyme product does not have an identical protein load *per se*, although the enzyme activity content may be the same. At first, a delivered enzyme product will contain a guaranteed minimal level of enzymatic activity. Since an enzyme undergoes some enzymatic inactivation upon storage between product generation and delivery, a certain overdose of enzymatic activity in the product will be used to guarantee a certain activity level. Secondly, each batch of enzyme produced will have a different history with respect to the processing efficiency and enzymatic inactivation profile during fermentation and downstream processing when the enzyme is (partially) purified.

5.4.2.9 Enzyme-containing wastewater

After enzymatic treatment, the medium drained off will contain protein, no matter whether the enzyme has been inactivated or not. Its presence in wastewater will lead to elevated levels of oxygen demand in municipal water purification units and often to increased costs of wastewater treatment if cost calculations are based on oxygen demand.

To limit the increase in wastewater treatment costs, the waste enzyme protein should be removed. This can be realised by various methods. The solubility of a protein in water is strongly reduced at its iso-electric point, the pH value at which the sum of all charged amino acids is zero. Shifting the pH of the wastewater to that pH value will initiate flocculation of some proteins. After complete precipitation, the protein can be separated by gauze filtration. Other enzyme proteins may be made insoluble by denaturation with temperature and/or pH shift. When dealing with enzymes that are not precipitated by one of these methods, the protein may be bound to resins prior to drainage. Separation of intact enzyme from the wastewater using membrane technology is another alternative, but will be more expensive and may be not economical. The separated protein itself may be used as a feed ingredient.

5.5 Health and safety issues

Large-scale application of enzymes started in the late 1960s when the first enzyme (protease) was incorporated in cleaning applications. In the textile production industry broad enzyme application was first used in the desizing process. Initially, companies working with enzymes underestimated their potency in relation to health. During this initial period of broad enzyme application, handling of enzyme (protease) preparations frequently resulted in skin irritation of employees. Employees came in direct contact with enzymes during production (fermentation), downstream processing and formulation of the enzyme product and in its application. Since then constant attention has been paid towards the risks to health when handling enzymes. This ongoing attention has resulted in, among others, application of formulation technology for enzyme preparations, improved industrial hygiene practices and procedures, improved design of production and manufacturing plant and dust control equipment, and increased understanding of enzyme-induced sensitisation and health effects. For the detergent industry it can be stated that, as a result of all the above-mentioned improvements and strict maintenance of hygiene rules, detergents containing enzymes have been produced safely during the last 25 years without the problems faced in the starting period (The Soap and Detergent Association, 1995). Nevertheless, when the strict hygiene rules are not obeyed in practice, adverse effects cannot be excluded. As an example in the textile industry, an employee of a dye house developed cellulase-induced occupational asthma as a result of daily exposure to this enzyme in powder form for almost two years (Kim *et al.*, 1999).

5.5.1 Exposure routes and health risks

Exposure to enzymes may cause irritation and/or allergies either via inhalation or by skin and eye contact. These allergies are no different to other irritants like pollen and house dust. The human body recognises an enzyme as human-different material and starts production of allergic antibodies when enough material enters the body via inhalation of enzyme-containing aerosol. When these allergic antibodies are detected, the person in question is said to be sensitised and the presence of antibodies is an indication that the person has inhaled human-different material. Determination of these antibodies during regular medical control can be used as proof that enzyme material has been inhaled. Prolonged uptake of enzyme material via inhalation can lead to symptoms such as watery eyes, runny nose, scratchy throat and occupational asthma, as mentioned before. When the intake of enzyme material ceases, these symptoms should disappear. However, when the doses of enzyme exposure are extremely high, this may cause irritation of the respiratory tract, resulting in symptoms such as congestion, difficult breathing and sore throat. However, under normal circumstances, with adequate controls or rapid action at the first signs of enzyme exposure, such situations should not occur.

Enzyme–skin contact will not lead to a sensitised human body. Skin irritation may occur in a situation of exposure to high levels of, especially, proteases, since the human body is relatively sensitive to this type of enzyme. The irritation caused by proteases is characterised by a weeping, red glistening appearance on the skin surface, which can be painful. The use of gloves and protective coveralls when working with large amounts of enzymes, no matter whether in granular or dissolved form, will prevent skin–enzyme contact. When small quantities of enzyme come into contact with the skin, a normal washing procedure with excess water should be effective in avoiding skin irritation.

Enzyme contact in the eye may have the same effect as skin contact: irritation. Rinsing thoroughly with excess water will minimise the effect of enzyme contact. The use of protective glasses and/or face shields is recommended when working with enzymes.

5.5.2 First aid

In a situation where all safety precautions are routinely followed, no contact with the active component of an enzyme product will occur. However, an accident with enzyme contamination as a consequence can occur at any time. Since the enzyme is soluble in water, water should be always used for the complete removal of the contamination upon direct exposure. It is therefore highly recommended to have safety showers and eye wash stations available in every area where possible enzyme contamination can occur.

For the various enzyme exposures the following actions are recommended:

- Ingestion: rinse the mouth and throat with tap water (the acid environment of the stomach inactivates the enzyme);
- Skin contact: wash the skin with plenty of water; remove clothing in case of contamination;
- Eye contact: rinse the eyes thoroughly with water for a long time (more than 10 minutes);
- Inhalation: remove the employee from the exposure area; in the event that the symptoms of irritation or sensitisation remain (shortness of breath and coughing), call or transport the person to a doctor.

5.5.3 Product design and allergenic potential

Enzymes are proteins which can catalyse certain chemical reactions. External proteins not belonging to the human body may have epitopes on their surface (local three-dimensional surface structures constructed by a combination of certain amino acids) which are recognised by antibodies when entering the human body. Every enzyme which is commercially exploited has to be previously tested in a series of toxicity tests (including determination of the doses, expressed as g/kg body weight, that are lethal for rats in 50% of the cases) in order to clarify its human and environmental toxicity potential. This information is incorporated in the MSDS (Material Safety Data Sheet) which has to accompany each enzyme product delivery.

The primary route for enzyme exposure which can potentially lead to allergy is inhalation. Therefore, the formation of respirable aerosols should be prevented as far as possible. The physical form of the enzyme can greatly influence the potential for such aerosol formation. Enzyme aerosols may be in the form of liquid droplets, mists, solid particles or dusts. Powdered enzyme products give the largest risk of exposure since they are easily aerosolised. For over 25 years, enzyme manufactures have formulated their products using granulation technology in order to prevent enzyme aerosol formation and the granulated products now available have excellent low-dust and abrasion resistance. The enzyme is encapsulated in these granules and in this way its release into the air is prevented. However, care must be taken not to crush them as the enzyme powder inside could then be easily aerosolised.

Enzyme formulations are also available in a slurry or liquid form. Slurries are more viscous than liquids and both have a minimum potential for aerosol formation, although situations have to be avoided where the chance for aerosol formation is increased. Examples of such situations are high pressure cleaning with steam, air or water and when any type of mechanical agitation is applied to the enzyme-containing liquid. Additionally, in the case of liquid enzyme spills, enzyme dust generation from dried material has to be avoided by washing away the enzyme material with water (Enzyme Technical Association, 1995).

A relatively new technology to reduce the allergic potential of enzymes is the modification of allergic epitopes on the surface of active and correctly folded enzymes by protein engineering. By use of this promising technology, variants of a protease from *Bacillus lentus* have been created with reduced allergenicity. Reductions to one-fifth of the level induced by the original protease have been measured by determination of the amount of allergy-specific antibody (Novozymes, 2001). Indications are that further modifications can reduce the allergenicity to almost zero without introduction of novel allergic epitopes.

5.5.4 Precaution and protection issues

An enzyme batch delivery should be accompanied by an MSDS. This contains information on how to handle the enzyme preparation and what safety precautions are required to protect the employee. Both employee and supervisor should have access to the sheets at all times. Additionally, enzyme manufacturers have manuals and booklets available for enzyme users, which will guide them in handling their enzyme products in a safe way. In the manual from Novozymes for instance, all safety, protection, monitoring and medical aspects are discussed (Novozymes, 2001). Also the Association of Manufacturers of Fermentation Enzyme Products (AMFEP), a European industry association founded in 1977, has produced an information guide concerning the handling of microbial enzyme preparations (AMFEP, 1994).

Proper work practice procedures are important in controlling enzyme aerosols and these have to be used in conjunction with engineering controls and personal protection equipment. These procedures include appropriate management systems for maintaining the necessary high hygiene standards, clear responsibilities for daily management, and availability of accurate trained personnel in the fields of medicine, industrial hygiene and engineering. The process plant should be engineered in such a way that risk of potential enzyme aerosol contact is minimised. Good local exhaust ventilation of the plant and enclosures are most effective instruments and should be preferably used at locations where enzymes are added into the process, at material transfer points and in the packaging rooms (Enzyme Technical Association, 2000). The process equipment should be a closed system when possible. Proper work practice procedures also include accessibility of washing facilities, cleaning equipment and disposal containers. Enzyme-containing spills and machinery should be cleaned directly by means of a vacuum system equipped with a high efficiency particulate air (HEPA) filter. Cleaning has to be done without brushing and/or sweeping since this will generate aerosol. The use of high-pressure water, air or steam has to be avoided for the same reason.

Employees who are working with enzymes should have access to personal protective equipment to prevent enzyme contact. This should protect the employee in a situation where process design and work-practice procedures are not appropriate. In other situations, such as clean up of spills, it will be the primary control method. Respiratory protection is used when airborne enzyme cannot be sufficiently controlled to a safe level. There are three main classes of respirators available: air purifying, air supplied and self-contained breathing equipment. The first class is most common in, for example, detergent facilities. The air-purifying respirator uses a cartridge to remove contaminants from the air. For enzyme-containing aerosols, a

HEPA filter cartridge is recommended. A higher level of respiratory protection is realised when using the air supplied system. In this situation, clean air is supplied to the wearer via a supply host. The self-contained breathing equipment obtains the highest level of protection and greatest freedom of movement and is, therefore, used in emergency situations. Once again, in normal operating conditions in a well-ventilated plant, the routine wearing of respirators is not necessary.

As noted before, skin and eye contact with enzymes should be avoided in order to prevent irritation. If there is a potential for eye or skin contact, protective wear and clothing should be used. Effective protective equipment include glasses, face shields, gloves and coveralls, and their use should be limited to the working area only and laundered after use.

5.5.5 Air monitoring

A continuously-operating air monitoring programme should be established in order to evaluate or screen the potential for employee exposure to airborne enzymes. Air samples should not only be taken at places where the highest risk of aerosol formation occur, but also in low risk places to get a complete overview of airborne dust existence within the plant. Historically, high volume air sampling was necessary because of the low sensitivity of the existing evaluation methods. Nowadays, operational evaluating methods are able to detect lower quantities of enzyme, which allow smaller samples for analysis. The benefits of working with small volume samples are that the required pumps are easier to calibrate, personal samples can be collected, and that the pumps are lighter, smaller and more mobile and do not require an external energy source. Upon sampling, airborne enzyme is collected on filters and the amount of enzyme is assessed either by measurement of the enzymatic activity or by ELISA (enzyme linked immuno sorbent assay). The sensitivity of the analytical ELISA method is high, while that of the activity measurement is dependent on which enzyme activity has to be measured (variable sensitivity). On the other hand, the analysis time for activity measurement is relatively short (less than one hour) when compared to that of the ELISA test (up to 12 hours). Which method is preferred will depend on the level of airborne enzyme expected and the time in which an analysis outcome has to be available.

The existence of enzyme exposure guides is important in a monitoring programme. They provide references for exposure levels that do not adversely affect the health of employees in an industrial environment where enzymes are used. As an example of a guide level, the ACGIH (American Conference of Governmental Industrial Hygienists) has established a threshold upper limit value of 60 ng/m^3 air for the protease enzyme subtilisin (The Soap and Detergent Association, 1995). This value is a ceiling value and the industry strives to have airborne enzyme levels which are normally an order of magnitude below this ceiling value.

5.5.6 Medical monitoring

Apart from monitoring the airborne enzyme throughout the production plant, the employees of the plant should also be monitored for possible enzyme contact. An ideal situation is when an employee, before he or she starts to work with enzymes, is examined medically. Specific topics in this pre-examination are their history of allergies, asthma, eczema, smoking, previous chest diseases, acquisition of a pulmonary function and enzyme sensitivity baseline and an overview of common allergens present (skin prick test) to determine the atopy of the individual. During his or her working life with enzymes, the employee should be routinely monitored annually. Deviation from this scheme may be considered in situations where, as an example, a new enzyme is introduced or process modifications have been adapted. Items that should be reassessed are the lung (pulmonary) function and enzyme sensitivity. Additionally, smoking and any relevant history of illness symptoms should be evaluated since the previous medical examination. Finally, employees should inform management and medical experts of a situation of any respiratory problem upon which prompt medical evaluation can occur and sources of airborne enzyme exposure can be removed.

Acknowledgements

The author thanks Novozymes and Mrs A.L. Auterinen and Mr A. Krouwer, both from Genencor Int. B.V., for their cooperation and the supply of documents concerning the safety issues surrounding enzymes.

5.6 References

AMFEP (1994) *Guide to the Safe Handling of Microbial Enzyme Preparations*, Brussels.

Auterinen A.L. (2002) Genencor International, personal communication.

Enzoguard® Brochure (1999–2001) Genencor International, Rochester, New York.

Edelstein S.J. (1973) *Introductory Biochemistry*, San Francisco, Ca, USA, Holden-Day Inc.

Enzyme Technical Association (1995) *Working Safely with Enzymes*, ETA, Washington. (Booklet)

Enzyme Technical Association (2000) 'Safe handling of enzymes', *Textile Chem. Color. Am. Dyestuff Rep.*, **32** (1), 26–27.

Kaya F., Heitmann Jr. J.A. and Joyce T.W. (1995) 'Influence of surfactants on the enzymatic hydrolysis of xylan and cellulose', *Tappi J.*, **78** (10), 150–157.

Kim H.Y., Nahm D.H., Park H.S. and Choi D.C. (1999) 'Occupational asthma and IgE sensitization to cellulase in a textile industry worker', *Annal. Allergy, Asthma Immunol.*, **82** (2), 174–178.

Lange N.K. (1999) Novozymes, personal communication.

Lehninger A.L. (1975) *Biochemistry*, Worth Publishers, New York.

Lenting H.B.M. and Warmoeskerken M.M.C.G. (2001) 'Mechanism of interaction between cellulase action and applied shear force, an hypothesis', *J. Biotechnol.*, **89**, 217–226.

Misset O. (1993) 'Stability of industrial enzymes' in *Stability and Stabilisation of Enzymes*, Proceedings of an International Symposium held in Maastricht, The Netherlands, Amsterdam, Elsevier Science Publishers B.V., pp. 111–131.

Novozymes (1999) Product sheet BioPrep, B1194a-GB, Bagsvaerd, Denmark.

Novozymes (2001) *Manual for Handling Enzyme Products*, LUNA No. 2001-19307-01 Bagsvaerd, Denmark.

The Soap and Detergent Association (1995) *Work Practices for Handling Enzymes in the Detergent Industry*, New York.

Ueda M., Koo H., Wakida T. and Yoshimura Y. (1994) 'Cellulase treatment of cotton fabrics. II Inhibitory effect of surfactants on cellulase catalytic reaction', *Textile Res. J.*, **64** (10), 615–618.

Website Novozymes, the bioindustrial e-zine, www.novozymes.com, 2002.

6

Effluent treatment – Enzymes in activated sludge

JOHN BINKLEY
University of Manchester Institute of Science and Technology, UK

ANDREAS KANDELBAUER
Graz University of Technology, Austria

Synopsis

The following will be discussed in this chapter:

- Typical components of textile effluent waste streams: the soluble components are dyes, oxidants, dissolved salts, surfactants, starches, etc. Insoluble enzyme substrates will mainly be fibres such as cotton, flax, viscose, lyocell, wool, silk, polyester, acrylics, polyamides and others. Details of the types of components, which are considered to be hazardous waste, will also be considered.
- Possible synergistic effects of other effluent components on colour removal with activated sludge, together with inhibition of the colour removal process, for example by heavy metals and dissolved salts. The physical state of the dye is also considered to be important in dye degradation mechanisms, such as the degree of aggregation of the dye. This can largely depend upon pH and other factors and components within the effluent during treatment.
- The influence of pH, temperature and concentration of other components.
- Mechanistic studies of aerobic and anaerobic enzyme activity. Enzyme activity under extremes of temperature, pH and other such conditions, which promote such activity.
- Aerobic and anaerobic dye degradation mechanisms that are likely to be in operation during treatment. Both are important and pertinent in an effluent treatment plant and the degree to which each predominates will depend on the rate and degree of oxygenation. Variation in degree of agitation within a plant, for example, will promote the formation of localised deadspots in which reduced oxygenation might favour a reductive dye degradation mechanism.
- Likely future trends encompassing minimisation of waste components. In the case of dyes, improved reactive dye bath exhaustion rates will be of importance.

- Alternative technology involving coagulation, flocculation and precipitation techniques.
- The current demand for stable dye systems so that resistance to dye degradation inherent within the dye structure. However, with increasing environmental awareness, there may be a tendency to move away from increased dye fastness especially on substrates, which have a limited service life.

  A book edited by Cooper (1995a) has significant contributions from other authors, many from different parts of the industrial sector.

6.1 Hazardous waste

Hazardous waste includes material described by the two groups shown below. Such wastes can be solids, sludges, liquids or gases.

- **Manifestation** – irritants, strong sensitisers, infectious, carcinogenic, mutagenic, teratogenic.
- **External influence** – flammable, explosive, radioactive, toxic or poisonous, corrosive.

Hazardous wastes include major types of heavy metals used in textiles and dyestuffs. Other lesser textile wastes – especially from dyeing and finishing – are acids and alkalis, bleaches, adhesives and polymers, crosslinking agents, carbonising agents, conditioner for wool, catalysts, detergents, dye carriers, chemical finishes including flame retardants and solvents. Of these, dye carriers, chemical finishes, and solvents are chemicals most likely to be potentially hazardous.

6.1.1 Heavy metals

These are potential toxicants dependent on their concentration and chemical form in the environment. Note that although arsenic is not a metal, it is included in this data because of its toxic nature. Table 6.1 gives metals frequently found in textile waste streams:

Processes and process materials in which these metals are commonly used are as follows:

- premetalised dyes with 3 to 4% metal content;
- double salt preparation of zinc in some basic dyes with approximately 3% metal content;
- dichromates used to oxidise and fix some dyes;
- chromium compounds as used in topchroming;
- catalysts for wash and wear applications;

Table 6.1 Metals frequently found in coloured textiles

Metal	[M] ppm	Dye type with highest [M]
Arsenic	<1–1.4	fibre reactive
Cadmium	<1	all types
Cobalt	3–83	acid
Copper	33–110	vat
Lead	6–52	fibre reactive
Mercury	0.5–1	vat
Zinc	3–32	basic

[M] refers to the concentration of the metal in the waste stream in ppm.

- improvement agents for wash and lightfastness in certain fabrics;
- flame-retardants.

6.1.2 Dyestuffs

Dyes tend to be non-biodegradable by conventional aerobic treatment systems. Some contain heavy metals such as Cr, Cu and Zn. Commercial dyestuffs comprise approximately 50% dye by weight, the remainder usually being non-hazardous filler, for example sugar and surfactant. Dye molecules contain a chromophore and an auxochrome, and the very position of substitution of the auxochrome can markedly affect the toxicity of the dyestuff. Many dyes contain heterocyclic compounds, which may exhibit chelating action. Therefore, toxicity could result, either by the removal of metals essential to the environment or by synergistic action to increase the toxic effects of metals normally present. Generally, from a dye class standpoint, basic dyes appear to be the most toxic because of their cationic nature. Direct and vat dyes tend to be non-hazardous.

From bioassay studies, basic (cationic) dyes and some acid and disperse dyes are toxic to fish and algae. The remaining dye classes are refractory, but may degrade anaerobically, typically in landfill sites, resulting in possible carcinogenic metabolites. Thus all dye-containing waste streams may be considered to be potentially hazardous.

6.2 Types of textile effluent

The scouring of cotton generates quite high concentrations of organic matter from natural impurities and from sizing materials in the fabric. This effluent tends to be concentrated, and therefore needs to be diluted prior to treatment, which also prevents clogging of the treatment plant.

Scouring synthetic materials yields effluent with appreciable amounts of size and antistatic agents in cloth. Raw wool gives high concentrations of emulsified wool grease resulting in highly polluted liquor.

In the bleaching process, the contents of bleach liquor effluent are mainly residues of oxidising agents, for example sodium hypochlorite ($NaClO$), sodium chlorite ($NaClO_2$) and hydrogen peroxide (H_2O_2), and alkalis or acids, which must be neutralised. Peroxides are not too troublesome but other oxidising agents must be reduced or kept out of the treatment plant altogether.

In mercerisation where caustic solutions are concentrated enough for processing, these can be further concentrated later by evaporation if necessary, when it may be economically viable to recover them. The effluent is a solution and suspension which is usually highly caustic and could be put to good use at minimal cost. For example, it could be mixed with acidic industrial effluent which is frequently high in heavy metals in order to precipitate out the metal hydroxides.

In dyeing, different types of dyeing processes will give rise to various kinds of liquor which may be discharged, and most or all will contain surface active agents. In sulphur dyeing, the liquor contains sodium sulphide, which can result in staining. This has to be removed or at least diluted. In chrome dyeing, liquors and some after-treatment solutions contain metals that can interfere in biochemical processes. These must be removed or made innocuous.

Printing wastes contain thickeners and dye pigment. There are five possible sources of these:

- Pastes surplus to requirements: these can be minimised by precise calculations in making up pastes.
- Residues on machines, screens and rollers: washings can be treated separately by sedimentation.
- Loose colour on fabric.
- Finishing residues from various starch fillings usually from surplus materials in mixing and machine boxes.
- Weaving size residues from washing out becks and sow boxes. Water jet looms utilise appreciable quantities of wastewater.

Corrosion in soft water at low pH seriously affects tanks and pipe work. Natural water and free CO_2 with no alkali or dissolved O_2 attacks iron pipes resulting in $FeCO_3$. Raising the pH with alkali prevents both types. Foaming was a serious problem in the 1950s. Detergents and soaps are now biodegradable. Unpleasant odours usually arise from organic wastes and solvents. The remedy for this is usually the use of flocculation, filtration and sedimentation techniques. We can also use adsorption onto active carbon or chlorination.

6.3 Methods of water treatment for incoming water

Clarification removes suspended solids, thereby making the water less turbid, and surface waters are the main type requiring this treatment. The exact level of treatment depends on the source; for example whether from river, lake, canal and so on.

The demand for water in printing, dyeing and finishing is usually quite variable, owing to the number of different operations that can be taking place at any one time. A system is therefore required which allows storage of both raw and treated water.

When demands are made on the treated water, it is replenished from the raw supply automatically, after having been treated first. Water can be drawn off at several depths, allowing take-off at or near the surface regardless of the level and ensuring that maximum settling has occurred, this being the first stage in clarification. Therefore a greater storage capacity means that a longer time for settling is necessary. Very small suspended particles will adsorb ions from the water and thereby acquire a charge. A charge layer will then exist around each particle, preventing coalescence. This charged layer therefore stabilises the particles and produces a finely divided suspension which, for many applications in wet processing is undesirable. Fine suspended solids may be removed by one of the following processes:

1 Coagulation, which is brought about by neutralisation of surface charge.
2 Flocculation, which is micelle formation by addition of a suitable flocculent.
3 Sedimentation, which is the settling of flocculated solids.
4 Filtration, which filters out flocculated solids.

Other substances, when present in significant concentrations, may be toxic. Volumes of wastewater and pollution loads are costly to treat; the lower the volume, the lower the cost of treatment. The volume is an important factor in determining the size of treatment plant. River authorities (now the Environment Agency in the UK) have high standards and because these require a high standard of treatment, it pays to reduce the volume.

6.4 Treatment of wastewaters from the textile industry

When they reach the manufacturer, cotton fibres contain natural waxes and fats which must be removed before bleaching or colouring matter is introduced to dyed yarn or fabrics. Scouring or kiering removes most of these.

Table 6.2 Typical scouring waste liquor from cotton linters

Properties and constituents	Concentration (mg/dm³)
Permanganate value (4 hour)	8850
BOD (5 day)	10650
Ammoniacal nitrogen	52
Total nitrogen	492
Phosphate	35
Potassium	148
Alkalinity ($CaCO_3$)	16750

Woven cotton fabric or yarn in the raw state is boiled in solutions of alkali, often under pressure.

Impurities possessed by woven cotton goods include natural impurities like those mentioned above. Size and starch, or various substitutes, are used to facilitate weaving. These can be removed by hydrolysis with acids or enzymes, followed by rinsing, or may be removed during the scouring process.

Scouring liquors tend to be alkaline and contain large amounts of organic matter (see Table 6.2), which are both expensive to treat. The standards of the river authorities, for example the Environment Agency in the UK, are very high.

After scouring, the cotton is bleached using:

$$\text{Hypochlorite, } HClO \rightarrow H^+ + ClO^-$$

Usually, potassium or calcium chlorite, salts or peroxide are used in bleaching. These remove and destroy natural colouring matter. Wastewaters from this process contain partly spent bleaching solution and impurities removed from the fibres. After bleaching, cotton is washed and then immersed in a solution of sodium bicarbonate or weak H_2SO_4, which destroys the bleach. Finally, a thorough wash with a soap solution is necessary.

Mercerisation increases lustre and dye affinity. The process involves treating the cotton fabric with sodium hydroxide, NaOH, and then washing with hot water whilst it is under tension. Any residual sodium hydroxide is neutralised with acid and the cotton material is thoroughly washed.

Dyeing of fabric results in large volumes of liquor being discharged. These liquors vary in character, depending on the nature and class of dye used; for example they may be acidic or alkaline and will contain high concentrations of salt and synthetic surface-active agents. Post-dyeing treatments involve the use of a variety of finishes such as starches, gums and waxes, resins and other materials, depending on the required finish, although these tend to be low in concentration because of the large volumes of water that are added during the finishing processes.

Wastewaters from the processing of wool include scour, bleach and dye liquors, plus associated washing waters. The two main differences from cotton wastes are the degree of alkalinity and the nature of the organic matter. Also, waste liquor from wool treatment may be acidic. Scouring wastewaters contain a high proportion of wool grease (lanolin), in an emulsified form, soluble organic matter and sand. Lanolin contains cholesterol and other complex organic substances, which are readily absorbed by the skin and are used in ointments and cosmetics.

Back washing liquor and yarn scouring liquor contain emulsified oils, grease and fatty matter derived from soaps. Emulsions must be centrifuged and then cracked with acid or other chemicals. Cracking is the process of breaking down into smaller molecules which are more amenable to further treatment processes. This aids wastewater treatment by making the soluble components more able to be broken down by biological treatment.

Greasy solids are normally removed by flocculation and filtration prior to mixing with other wastewater. Mixing and balancing waste liquors evens out large variations in composition, temperature and flow. Balancing involves neutralising acid or alkali so that the liquid is in a suitable condition to go forward for secondary treatment.

Waste liquor from wool treatment is similar to that from cotton dyeing but may be acidic. It contains large amounts of wetting agents and softeners plus some salts and potentially toxic materials. These can be inhibitory to biochemical processes, i.e. they retard or tend to prevent naturally occurring biological processes. Waste liquor from the treatment of synthetic fibres contains a variety of organic substances at several hundreds of milligrams per litre many only slowly degraded by special microorganisms.

6.5 Effluent treatment

The methods of treatment available are the same whether the effluent is treated on site or at the local sewage treatment works (STW). The various methods of treatment can be categorised under the following headings:

- membrane technologies.
- chemical treatment
- biological treatment

The way in which these treatments work is discussed in the following sections.

6.5.1 Membrane technologies

A variety of filters are used to remove the contaminants from textile effluent and these include the following techniques:

- Reverse osmosis is a high-cost treatment where the effluent, under pressure, passes across a semipermeable membrane. A purified permeate of relatively clean effluent and a concentrate of the contaminants are formed which have to be disposed of in some other way. The permeate will still contain some residue of contaminants as approximately 20% of the effluent remains untreated. The membrane needs frequent cleaning which results in additional cost and a more labour intensive process (Cooper, 1995a).
- Nanofiltration involves the use of a membrane as a filter. It retains any materials with a relative molecular mass greater than approximately 200. The permeate may still contain high levels of dissolved inorganic salts and therefore cannot be used for recycling. The concentrate contains all the organic impurities and some of the inorganic ones and will subsequently need further treatment. The need for frequent cleaning, together with high capital costs and high running costs make the units very expensive to run, even more so than reverse osmosis (Cooper, 1995a).
- Ultrafiltration or microfiltration aims to reduce suspended solids and organic materials with a particle size of 0.02 mm or greater. These solids form sludge, which is subsequently disposed of. They have little effect on the reduction of colour and any colour that is removed is adsorbed onto the other materials during removal (Cooper, 1995a). Both microfiltration and ultrafiltration are suitable for reducing chemical oxygen demand (COD) and suspended solids but their use would have to be as part of a multitechnique treatment plant.

6.5.2 Chemical treatment

Chemical treatment is usually simply coagulation and flocculation. Under optimum conditions only approximately 50–60% biological oxygen demand (BOD) (organic matter) can be removed by addition of coagulants such as $Al_2(SO_4)_3$ or $Fe_2(SO_4)_3$ and lime. Therefore this process alone will not achieve the standards required by river authorities.

The method is well established as an effective and economic way of treating effluent. A separation process, i.e. sedimentation or flotation, is always used in conjunction with this process. Chemicals are used to form a precipitate, which during formation or settlement adsorbs the colour and other unwanted materials in the effluent. The precipitate is removed as a sludge, which will probably be dewatered to reduce the volume. In the past, inorganic coagulants such as lime, magnesium and iron salts were used to remove modern dyes but now organic polymers have been developed for more efficient treatment of the colour. These coagulants are not without problems and in some cases colour consent conditions are not met. Some polymers can cause toxicity problems in the discharging effluent. Many of the polymers used are cationic and cause toxicity problems in addition to adverse effects on nitrification in the discharging effluent, which could in turn affect any activated sludge process that the effluent subsequently passes through.

Chemical oxidation may be employed to remove colour from effluent. Strong oxidising agents such as hydrogen peroxide, chlorine, ozone etc. are used to degrade the organic molecules and the dye. This is however a costly treatment (Socha, 1992). Partial oxidation has been tried as an economically viable study but problems lie in the nature of the products formed, which could be more harmful to the environment than the constituents of the original effluent (Cooper, 1995a).

6.5.3 Biological treatment

Most materials of animal and vegetable origin can be broken down into simpler compounds such as water and carbon dioxide by microorganisms such as bacteria, fungi, protozoa and so on. This can be done in the presence of oxygen by aerobic degradation or in the absence of oxygen by anaerobic degradation. Both happen as the bacteria and other microorganisms convert the effluent into more acceptable products (Socha, 1992). It is important that the type and concentration of substances, and conditions such as pH, temperature and so on that will kill off the microorganism population are carefully controlled.

Aerobic treatment can be carried out in stabilisation ponds, aerated lagoons, activated sludge or percolating filters. Aerobic treatment uses oxygen dissolved in the wastewater together with microorganisms in the activated sludge to convert the wastes to more microorganisms and CO_2.

Organic matter is partially oxidised and some of the energy produced is used for making new cells with formation of flocs. The flocs are allowed to settle and then removed as sludge (Laing, 1991). A proportion of the sludge removed is recycled back to the aeration tank to maintain the colony of microorganisms and the remainder of the sludge can either be disposed of or further reduced by anaerobic treatment. Disposal of the sludge can be through agricultural use as a fertiliser, in landfill or by drying and incineration (although, disposal to agriculture is prohibited by law in many countries because of the presence of heavy metals).

Anaerobic treatment occurs in sealed tanks and converts the waste into methane and carbon dioxide. Where nitrogenous and sulphide-containing pollutants are present, ammoniacal substances and hydrogen sulphide are

Table 6.3 Conditions necessary to ensure no inhibition of biological treatment (Laing, 1991)

Variable	Condition
Temperature	35°C maximum (can be higher for anaerobic treatment)
Ratio BOD:nitrogen	Approx. 17:1
Ratio BOD:phosphorus	Approx. 100:1
pH	6.5–9 (preferably 7 maximum)
Metals (Zn, Cu, Cr)	Less than 10 mg/dm^3

produced. At some municipal sewage treatment works, the sludge formed by the aerobic treatment process passes into tanks for anaerobic treatment. Considerable heat is produced from anaerobic treatment and after extraction by heat exchangers can be used to heat administration buildings within the sewage treatment works (STW). The methane produced is collected, compressed and then used in generators to produce electricity. The electricity produced can power site processes and the surplus is sold to the national grid. The production of this power not only reduces the running costs of the STW but also provides a welcome income, thus reducing costs further.

For biological treatment to succeed there must be sufficient nitrogen and phosphorous present in the effluent. It may be necessary to add these nutrients in an industrial treatment plant. At a STW where domestic effluent is mixed with industrial effluent, these nutrients should appear in high enough concentrations to keep the microorganism population healthy. Table 6.3 shows other conditions that are necessary to ensure the success of biological treatment.

6.6 The use of activated sludge for the removal of colour

The extent of the problem posed by colour depends very much on the class of dye that is used. Some dyes are very much more easily removed from effluent than others. Insoluble vat and disperse dyes can be removed in quite high proportions by primary settlement. Basic and direct dyes respond well to treatment by the activated sludge process but reactive dyes and some acid dyes seem to cause more of a problem. It is generally considered that the activated sludge process removes only low levels of these dyes. Some dyes will respond better to anaerobic conditions than aerobic conditions (Paul *et al.*, 1997). Many dyes are not biodegraded but adsorbed under aerobic conditions. Studies have found that many azo dyes can be degraded under anaerobic conditions (Cooper, 1995a). Twenty-two water soluble dyes were assessed for degradability under anaerobic conditions. Results showed that all except one of the anthraquinone acid dyes, CI Acid Blue 80, showed signs of significant colour removal.

Conclusions were drawn and suggestions made that these dyes were likely to be broken down in the environment by anaerobic conditions. Concerns were raised that the aromatic amines formed during this process could prove more toxic to the aquatic environment than the intact dye molecule. ETAD (The Ecological and Toxicological Association of Dyes and Organic Pigment Manufacturers), which was formed in 1974 to represent the interests of these industries on matters relating to health and environment, investigated the anaerobic degradation of simple aromatic amines and found they were unlikely to be persistent (Cooper, 1995a). Trials were carried out in simulated landfill conditions using sludge contaminated by azo dyes and no evidence of dye in the leachate was found nor any amines that would occur as metabolic byproducts (Laing, 1991).

Today's consumers demand textiles that are fast, not only to washing but also to light and perspiration. Much research has produced dyes with structures that are much more stable to such conditions than in the past. This stability also makes microbial attack more difficult, limiting their biodegradability.

It has long been known that the ease of elimination of acid dyes from effluent was directly related to the solubility of the dyes (Hitz *et al.*, 1978). The more sulphonic acid groups that are present in the structure of the dye, the more soluble and therefore less responsive to treatment is the dye by the activated sludge process. The removal of acid dyes by bioelimination, which is adsorption onto the biomass rather than biodegradation, was found to be related, not to the number of sulphonic acid groups (and thereby the solubility), but to the size of the dye molecule. It is thought that the greater the molecular size, the greater the degree of adsorption (Cooper, 1995b).

Early colour removal researchers found in their study that medium to high rates of colour removal were obtained for the disperse dyes (Hitz *et al.*, 1978). The removal of water-soluble basic and direct dyes is good although this is probably by adsorption onto the biomass rather than biodegradation. The reactive dyes showed little evidence of colour removal, which bore no relationship to the number of sulphonic acid groups or the ease of hydrolysis. No vat dyes were used in this study because of the more complicated procedures needed to prepare the insoluble vat dyes. It would be expected that much of the insoluble disperse and vat dyes could be removed by primary sedimentation prior to any treatment. It was found in this study (Hitz *et al.*, 1978) that reactive dyes respond poorly to the activated sludge process.

Adsorption onto the biomass is poor, with a maximum of 30% colour removal, and biodegradation is virtually non-existent.

In a more recent study (Binkley *et al.*, 1998), different results were obtained. It was found that using an activated sludge obtained from a conditioned textile effluent treatment plant, a removal rate well in excess of the 30% previously mentioned was obtained. In this study removal rates well in excess of 80% were found using the conditioned activated sludge.

A broad range of microorganisms is needed in the activated sludge process to treat the diverse range of pollutants found in industrial and domestic effluent that passes through the STW prior to discharge to the watercourse. At the STW the constituents of the incoming wastewater can change considerably on an hourly or daily basis.

The industrial on-site effluent treatment plant has a more consistent quality to its effluent. The range of pollutants is much more specific to particular industrial processes and it is therefore known on a daily basis what the effluent will contain. The diversity of the microorganism population is probably much less than that in a municipal STW and therefore more specialised. As a result, the sludge will be conditioned to the chemicals normally found in the effluent. Consequently, any shock loading with foreign pollutants can severely hamper the smooth running of the plant. This is also the case for municipal STW plants.

The precise composition of diverse microorganisms used in the treatment of wastewater is difficult to define. A complete ecosystem is developed which feeds on components of the incoming effluent and each other. The composition of this ecosystem will depend on the environment created by such conditions as pH, temperature and the availability and variety of the food (Horan, 1991).

The largest component of the population in this mix of microorganisms is the bacteria. Bacteria are very small, the majority having diameters of 0.2–1.5 μm and are very difficult to see using an optical microscope although a microscope with a good resolving power will allow identification of some of the shapes of the larger bacteria. Staining techniques, using a dye, can make the bacteria easier to see and identify, and the whole microorganism or just a selected part of it can be coloured.

Four different shapes of bacteria can be identified: sphere, straight rod, curved rod and spiral.

The sphere can be found singly (coccus), in pairs (diplococci) or in a chain (streptococci). The straight rods are the most commonly found bacteria and include the frequently seen genera such as *Pseudomonas*, *Zooglea*, *Eschericha* and *Salmonella*. The curved rods (vibrio) are a single curve whereas the spirals (spirilla) (normally only found in water) can vary from one complete turn to many turns and can measure from 0.5–60 μm in length (Horan, 1991).

Protozoa are also an important component of this microbial mix. They are single cell organisms which can be found in several forms. Some are well known as pathogens, i.e. they are agents of disease and death, but most are harmless. Some are parasitic but most live quite freely in all types of habitat. Protozoa prove to be very valuable in activated sludge treatment plants. They are more common in aerobic than in anaerobic treatment (Curds, 1992).

The larger protozoa are clearly visible under the microscope making identification a lot easier than the bacteria. Most common protozoa vary in size from 5–250 μm although some very rare ones have been found as large as 6 mm in length (Pechenik, 1991). Three types of protozoa can be found in wastewater treatment processes. These are:

- *Mastigophora* (flagellated protozoa), which possess one or more flagella that are used to assist in feeding and motility, giving them the ability to move spontaneously under their own power. They can be subdivided into two classes, *phytomastigophora* – algae fall into this class – and *zoomastigophera*, most commonly found in activated sludge plants.
- *Sarcodina* (amoeba), which have pseudopodial structures that are involved in feeding and the locomotion of the organism. Amoeba can be naked, lacking any skeletal structure or testate with shells composed of proteinaceous, silicaecous or carbonaceous material.
- *Ciliophora* (the ciliates), which are the largest in terms of species. A characteristic of these is the arrangement of cilia over the surface of the cell, used for purposes of locomotion. The ciliates can be subdivided into four groups:
  - i *holotrichia* – free swimming, cilia all over the body
  - ii *spirotrichia* – flattened body with cilia on the underside
  - iii *peritrichia* – inverted bell-shaped bodies, mounted on a stalk, with cilia arranged around the top of the body. The stalk anchors the organism to a sludge floc.
  - iv *suctoria* – cilia in early life; they develop a stalk and feeding tentacles later (Horan, 1991).

Some protozoa feed on bacteria and other organic matter while others prey on smaller protozoa and as such are considered predators. Also present in this microbial mix are rotifers and nematodes.

- Rotifers are simple multicellular invertebrate animals, 50–250 μm in length. They have a ring of cilia surrounding the mouth, which sweep bacteria and other organic matter in. Some rotifers swim but others move by a creeping motion. The body anchors itself to a floc and can be seen stretching out from the floc surface. They are quite widespread and are desirable for the two roles they play in an activated sludge process. Rotifers remove freely suspended bacteria, i.e. non-flocculated bacteria,

and they make a contribution to the floc-forming process by producing faecal pellets surrounded by mucus (Bitton, 1994).
- Nematodes are unsegmented cylindrical worms, 0.5–3 mm in length. They feed on bacteria and other microorganisms. They are often found in the activated sludge process but it is thought that they have little to do with the process of decreasing COD (Henz *et al.*, 1995).

Organic and inorganic particles together with bacterial cells form the constituents of an activated sludge floc. The size of flocs can range from less than 1 μm to 1000 μm (Bitton, 1994). The flocculation process brings together primary particles to form a floc. The primary particles are very small, too small to settle out and if unable to flocculate will remain as suspended solids in the settlement tanks. As the floc size increases, the number of active aerobic bacteria will decrease and anaerobic bacteria such as methanogens take over in the inner regions of the floc (Bitton, 1994). It has been suggested that the structure of an activated sludge floc can be explained as filamentous microorganisms forming the backbone of the floc to which zooglea and floc-forming microoganisms attach themselves. Some flocs do not have this filamentous backbone; therefore a different theory has been proposed suggesting that some of the activated sludge microorganisms produce polymers that are responsible for floc formation (Bitton, 1994). Although these are just two of the theories put forward, the exact composition and structure of the sludge flocs is uncertain (Henz *et al.*, 1995). Dense and fairly regular flocs with evidence of protozoa amongst them are a sign that the biomass is in a healthy state. Mistreatment of the activated sludge will result in the protozoa disappearing and disaggregation of the flocs into their primary particles. It is only when these primary particles are brought together that the flocs are formed and are able to settle out of the process for disposal. Some form of agitation is necessary to ensure that the primary particles meet up but the flow rate should not be high enough to inhibit settlement. In plants with a high flow rate, flocculation is poor, a high degree of suspended solids is found and turbidity of the effluent is evident. It therefore takes the skill of the operative to maintain the correct level of agitation.

6.7 Decolourisation – by enzymes, fungi, and by biosorption and enrichment cultures

6.7.1 Biotechnological methods

For microbial dye decolourisation not connected to the activated sludge process, bioreactors can be designed containing defined bacterial or fungal cultures, isolated enzymes or enzyme mixtures immobilised on a solid carrier material. Such immobilised biocatalysts are much more susceptible to damage by harsh chemical environments than are the mixed cultures present in activated sludge systems. Thus, they are preferentially used to treat partial process streams within the plant where the composition and physical properties are rather more readily controlled than the overall plant effluent.

For instance, exhausted dyeing bath solutions have been decolourised enzymatically using a laccase formulation and subsequently reused for the preparation of new dyeing baths (Abadulla *et al.*, 2000). Similarly, hydrogen peroxide has been removed enzymatically from bleaching solutions using a catalase in an industrial pilot experiment (Paar *et al.*, 2001). Owing to the use of high-specificity enzymes, only the target molecules are attacked while valuable additives or fibres are kept intact and can be reused.

6.7.2 Enzymes

Because enzymes are protein molecules they do not metabolise dyes but they do catalyse a specific type of transformation with them. Mineralisation of dyes can never be expected with enzymes. However, enzymatic modification of dyes may often be sufficient at a certain stages in the process to destroy chromophores and reduce toxicity. Additionally, enzymatic modification can render the dyes more susceptible to subsequent biodegradation in a municipal wastewater treatment plant where complete elimination takes place.

Although dye molecules display a high structural variety, they are degraded by only a few different enzymes. These biocatalysts have one common mechanistic feature: They are all redox-active molecules and thus exhibit the desired broad substrate specificities.

In the early 1980s, there were reports about a specific enzyme responsible for azo dye cleavage (Zimmermann *et al.*, 1982, Zimmermann *et al.*, 1984). This enzyme was called an azoreductase. The authors had isolated a bacterial strain capable of degrading the azo dye Orange II. Subsequently they succeeded in the isolation and characterisation of a single protein that catalysed the azo bond reduction with remarkable specificity towards the dye used for the long-term adaptation of the organism with remarkable specificity towards Orange II. The reduction of the azo compound took place in the presence of oxygen, which is especially noteable since normally anaerobic conditions are required.

Since then, a number of reports have been published where azo dye decolourisation has been linked to some azoreductase activity of the described organisms. However, the difficulties in finding organisms that can grow solely on azo dyes as carbon and energy sources showed that such azoreductase activity often is a secondary metabolic effect caused by the action of unspecific reductive enzymes such as cofactor-dependent oxidoreductases or cytochrome P450 reductases.

Moreover, azo dye reduction does not necessarily depend on the action of enzymes at all. They may simply be reduced by the reductive environment generated within an organism. Thus, cleavage of azo bonds under anaerobic conditions is often due to unspecific reduction processes mediated by redox-active compounds such as quinone-type substances, biochemical cofactors such as NADH or reduced inorganic compounds such as Fe^{2+} or H_2S which are formed by certain strictly anaerobic bacteria as metabolic end products. As a consequence, types of dyes, other than azodyes, that are susceptible to reduction can be transformed as well. In contrast, the azoreductase described by Zimmermann *et al.* (1984) displays very restricted substrate specificity and is limited to a very narrow range of structurally related azo dyes.

Generally, in the presence of oxygen, intracellular mono- and di-oxygenases that are ubiquitously present in living organisms may transform dyes in the course of secondary metabolism. They cause the breakdown of aromatic rings via incorporation of oxygen atoms (biohydroxylation) and subsequent cleavage of the ring system resulting in carboxylic acids, which are further used in metabolism (Fig. 6.1).

The enzymes mentioned so far are dependent on cofactors such as NAD(P)(H) or FAD(H). They are never used in isolated form but only within whole cell systems where the cofactors are regenerated continuously. For technical applications of single enzymes or enzyme cocktails less complicated systems are required.

In terms of enzyme remediation of textile dyes, laccases seem to be the most promising enzymes. Laccases are polyphenoloxidases which accept a wide range of aromatic alcohols and amines as substrates. Their major advantage lies in that there are no expensive cofactors required. Only oxygen (i.e. air) needs to be present as a cosubstrate. Laccases have been shown to decolourise a wide range of industrial dyes and in the presence of small amounts of low molecular weight redox mediators their versatility can be extended even more (Reyes *et al.*, 1999). Not only azo dyes are attacked by laccases. Various other types of dyes are also attacked. For example, indigoid dyes are decolourised very efficiently by laccases from various fungi. As shown in Fig. 6.2, indigo (1), the most important dye in the manufacturing of blue jeans, was demonstrated to be cleaved under laccase catalysed electron transfer to give isatin (2) and upon further decarboxylation anthranilic acid (4) as the final stable oxidation product. It was suggested that the degradation might proceed via dehydroindigo as a reaction intermediate (Campos *et al.*, 2001). This process is used industrially to achieve the stonewashed effect of indigo-dyed denim fabric by means of milde enzymatic decolourisation.

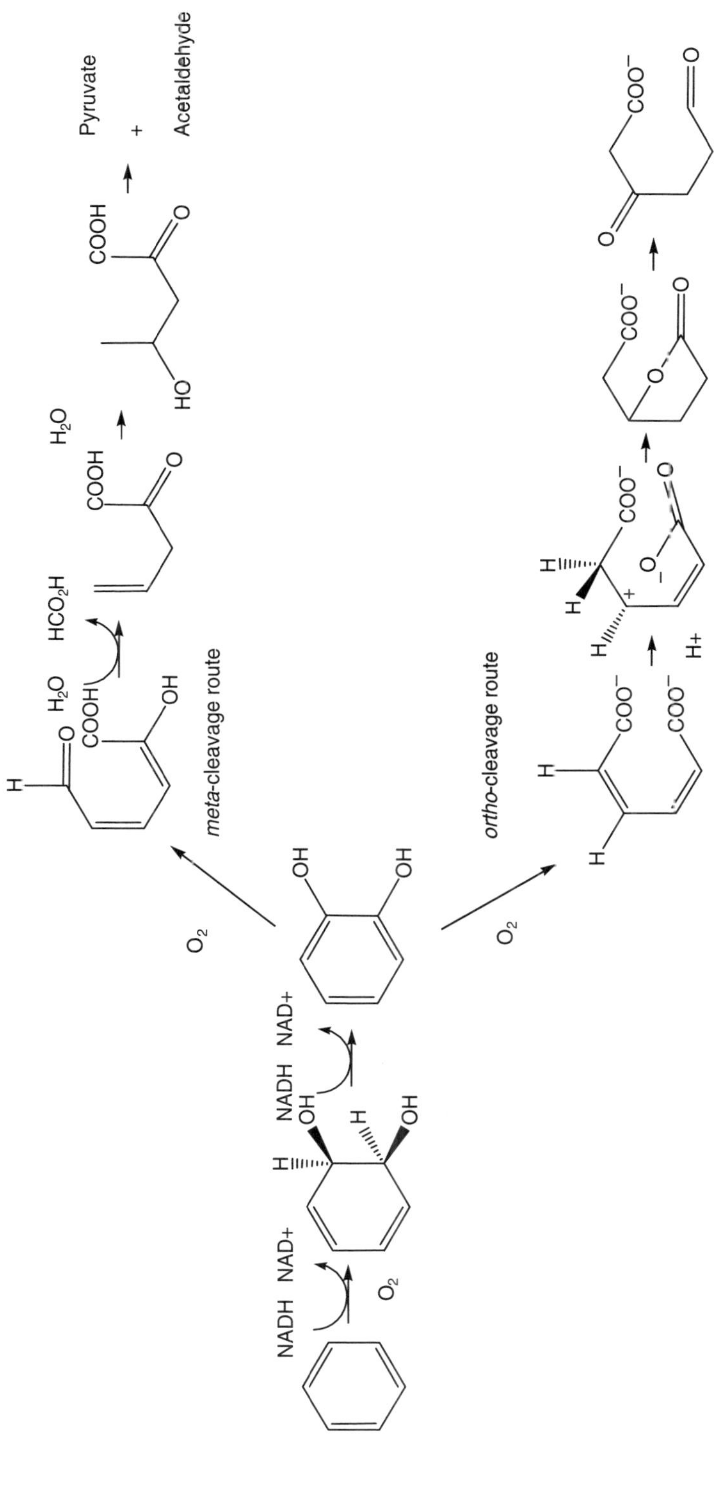

6.1 Enzymatic oxidative cleavage of aromatic rings.

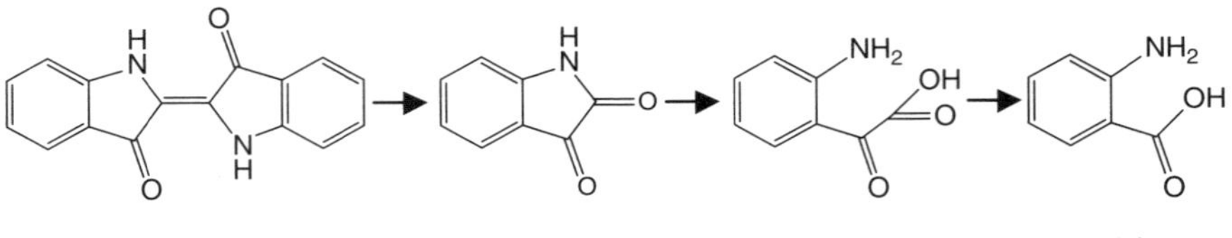

6.2 Suggested oxidative degradation pathway for indigoid dyes (Campos *et al.*, 2001).

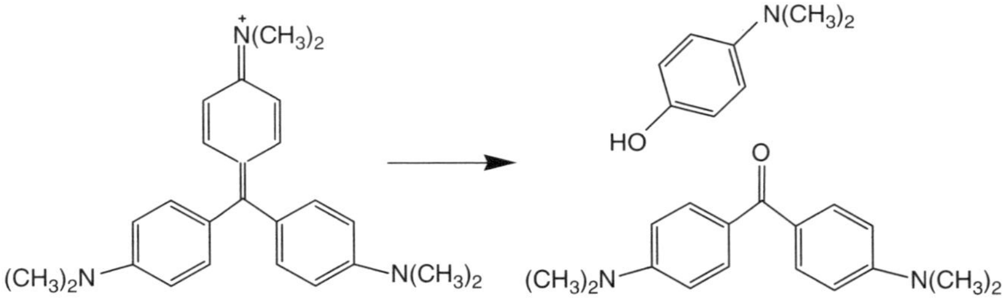

6.3 Decolourisation products suggested for triarylmethane dyes (Yatome *et al.*, 1993).

Another important enzyme type involved in the decolourisation of textile dyes is the class of peroxidases. Like laccases, they do not require expensive cofactors but use hydrogen peroxide as an auxiliary for the oxidation of aromatic compounds. As discussed previously, in the bacterial systems present in activated sludge they catalyse the oxidative cleavage of azo dyes. Again, other classes of dyes are susceptible to peroxidase catalysed oxidation as well. Triphenyl methane dyes for instance, which are especially recalcitrant towards biodegradation, can be oxidised by peroxidases (Fig. 6.3). Substrate specificities were thereby shown to differ from those observed with laccases (Shin and Kim, 1998). With both enzymes, anthraquinoid-type dyes can be dealt with as well.

Oxidases from various different kinds of organism are readily available and display varying substrate specificities, inhibition characteristics and pH- and temperature profiles. For example, fungal laccases are well suited for applications in acidic solutions down to pH 3, while they usually are inactivated at pHs above 7 to 8. Bacterial laccases, on the other hand, display their optimum activity in the alkalic region and consequently may be employed to complement fungal enzymes. Similarly, enzyme biocatalysts can be tailored for specific applications depending on the type of predominant contaminant, additive environment or temperature range.

6.7.3 Fungal decolourisation

Both types of oxidative enzymes described previously, laccases and peroxidases, are predominantly generated by white-rot fungi (one of their natural functions being to degrade lignin), a complex aromatic matrix in wood. The fact that laccases and peroxidases are excreted by the fungi makes the latter organisms themselves interesting for application in bioremediation. As living whole-cell systems they are suitable for use as end-of-the-pipe solutions as an alternative to activated sludge processes. Various white-rot fungi like *Phanerochaete chrysosporium*, *Trametes versicolor* or *Pleurotus ostreatus* have been employed, immobilised on various supports. Fluidised-bed reactors, for instance, which utilise fungi, contain freely-mobile pellets covered with layers of immobilised biomass, while packed-bed reactors contain organisms that are fixed onto a suitable support material (Zhang and Yu, 2000). Packed-column reactors and rotating drum reactors using fungal biofilms have also been constructed (Kapdan *et al.*, 2000). In all cases, the fungal mycelia are grown on the solid support directly in the reactor. By addition of various types of nutrient, the efficiency of the systems can be manipulated. For an informative compendium of the literature describing fungi used to decolourise dyes, see the review of Fu and Viraraghavan (2001).

Although both immobilized fungi and isolated enzymes have a good potential for the treatment of process waters in the textile industry, enzyme reactors display one major advantage over whole-cell systems; with the latter, enzymes are produced *in situ*. Thus, the concentration of the active compound, which is always an enzyme, is limited by the growth of the living organism. In contrast, with isolated enzymes appropriate amounts can simply be dispensed. To support fungal growth, nutritional compounds must be added while in parallel other substances are generated in addition to the desired biocatalyst. This may lead to unwanted side-effects and, at least, will incorporate new loadings of chemicals into the effluent. Furthermore, enzyme expression and thus its secretion by a fungus is dependent on its metabolic state and therefore presents far from constant release into the medium over time.

6.7.4 Decolourisation by biosorption

A prominent method for removing colour from effluents is to adsorb coloured particles physically onto various materials like charcoals, clays, soils, diatomaceous earth, activated sludge, compost, living plant communities, synthetic polymers or inorganic salt coagulants (Slokar and Marechal, 1998). If biomass is used for such an adsorption, the process is called biosorption.

Colour removal via biosorption is usually achieved by adsorption on fungal mycelia. Fungal cells for biosorption applications may either be used as growing cells or in the form of dead biomass; thus decolourisation can take place with or without concomitant biodegradation. Not surprisingly, decolourisation with active biomass is usually much more effective owing to parallel digestion (Aretxaga *et al.*, 2001). In the former case, the problem of elimination of waste dyestuff is not actually solved since the recalcitrant molecules are still present and have to be treated in a subsequent step.

Azo dyes have been shown to bind effectively onto the mycelium of *Aspergillus niger* resulting in extensive colour removal at greater than 95% (Sumathi and Manju, 2000). A stationary culture of this fungus was also used to decolourise a complex wastewater from a textile company by an airlift bioreactor over a relatively wide range of pH values. Between pH 3 and 7 there was 100% decolourisation; with pH 12 still about 60%. Thus, the process displays a high tolerance towards harsh changes in the pH. Furthermore, it does not seem to be limited to a certain type of dye. Acid, basic, direct, reactive and disperse dyes are reported to be removed from solution within a couple of hours (Assadi and Jahangiri, 2001).

Pellets consisting of activated carbon and mycelium of *Trametes versicolor* were used for textile dye decolourisation by Zhang and Yu (2000). Combining biodegradation with adsorption, high decolourisation rates could be achieved as was also reported for a system using bacteria and carbon black as a carrier material (Walker and Weatherly, 1999).

Biosorption on agricultural residues was suggested by Nigam and coworkers as a first step prior in microbial treatment to concentrate dyes (Robinson *et al.*, 2001). Here, dyestuff is first adsorbed on a waste product that later serves as a growth substrate for solid-state fermentation of a suitable fungus. This seems to be a very promising alternative, since in a first step, very rapid decolourisation of the effluent takes place by physical means compared to the rather slow decolourisation by growing organisms which may take up to a couple of days until satisfactory decolourisation is achieved. This slow step of fermentation of dyes is carried out separately, when time plays a less significant role. Moreover, the growth conditions and thus the decolourisation activities are optimal for white-rot fungi during solid-state fermentation. Comparable decolourisation rates could never be achieved with the fungi growing in the effluent.

6.7.5 Enrichment cultures

Enrichment cultures are populations of microorganisms or single organisms which have developed a certain property via natural adaptation. In general, enrichment of microorganisms with special effectiveness in dye digestion occurs at any site where dyes are present in large amounts. Dye decolourising microorganisms can hence be obtained simply by isolation of existing dye-degrading cultures from environmental samples such as textile effluents.

Applying methods of directed evolution, adaptation of promising strains to the conditions present in textile effluents is possible as well. Under chemostat (an apparatus for growing bacterial cultures at a constant rate by controlling the supply of nutrient medium) conditions, enrichment of bacteria capable of growing on dye molecules as the only carbon source takes very long periods of time; from several months up to years. Alternatively, employing genetic methods, DNA encoding for enzymes involved in dye metabolisation can be transferred and new organisms with desired properties can be tailormade.

Several enrichment cultures have been described which have been optimised for special requirements. For instance, thermophilic bacteria, selected by adaptation from a textile effluent, have been shown to decolourise textile dyes at temperatures up to 60°C. One isolate of this mixed culture was able to decolourise commercial azo, diazo, reactive and disperse dyes (Banat *et al*., 1997).

Complete mineralisation of dyes under various conditions by using enrichment cultures has also been accomplished. A methanogenic consortium was found to detoxify aromatic amines formed during the prior azo reduction step resulting in the complete mineralisation of an azo dye under strict anaerobic conditions. This mixed population was grown on the amines as the sole nitrogen sources (Razo-Flores *et al*., 1997) and the azo dye azodisalicylate was continuously degraded in a bioreactor for more than 100 days.

Complete mineralisation of an azo compound by an isolated aerobic bacterial strain has also been successful. Via continuous adaptation of *Hydrogenophaga palleronii* S1, a strain was developed growing on 4-carboxy-4′-sulphoazobenzene as the sole source of carbon and energy (Blümel *et al*., 1998).

6.8 References

Abadulla E., Tzanov T., Costa S., Robra K.H., Cavaco-Paulo A. and Gübitz G.M. (2000) 'Decolorization and detoxification of textile dyes with a laccase from *Trametes hirsuta*', *Appl. Environ. Microbiol.* **66**, 3357–3362.

Aretxaga A., Romero S., Sarr M. and Vicent T. (2001) 'Adsorption step in the biological degradation of a textile dye', *Biotechnol. Prog.* **17**, 664–668.

Assadi M.M. and Jahangiri M.R. (2001) 'Textile wastewater treatment by *Aspergillus niger*', *Desalination*, **141**, 1–6.

Banat I.M., Nigam P., McMullan G. and Marchant R. (1997) 'The isolation of thermophilic bacterial cultures capable of textile dyes decolorization', *Environ. Int.* **23**, 547–551.

Binkley J., Hargreaves J., Evans T.M. and Smart G. (1998) 'Biochemical treatment of recalcitrant dyestuff effluent', *Ecotextile 98, Proceedings of the 2nd International.* Textile Environment Conference, Bolton, UK. 7 & 8 April 1998, p. 66.

Bitton G. (1994) *Wastewater Microbiology*, Wiley-Liss, New York.

Blümel S., Contzen M., Lutz M., Stolz A. and Knackmuss H.J. (1998) 'Isolation of a bacterial strain with the ability to utilize the sulfonated azo compound 4-carboxy 4′-sulfoazobenzene as the sole source of carbon and energy', *Appl. Environ. Microbiol.*, **64**, 2315–2317.

Campos R., Kandelbauer A., Robra K.H., Cavaco-Paulo A. and Gübitz G.M. (2001) 'Indigo degradation with purified laccases from *Trametes hirsuta* and *Sclerotium rolfsii*', *J. Biotechnol.*, **89**, 131–139.

Cooper P. (ed). (1995a) *Colour in Dyehouse Effluent*, Society of Dyers and Colourists, Bradford, UK.

Cooper P. (1995b) J. Textile Inst., **84** (4), 553.

Curds R. (1992) *Protozoa in the Water Industry*, Cambridge.

Fu Y. and Viraraghavan T. (2001) 'Fungal decolorization of dye wastewaters: a review', *Biores. Technol.* **79**, 251–262.

Henz M. *et al.* (1995) *Wastewater Treatment*, Springer.

Hitz H., Huber W. and Reed R. (1978) *J. Soci. Dyers. Colour.*, **94** (2), 71.

Horan, J. (1991) *Biological Wastewater Treatment Systems*, J. Wiley and Sons, London.

Kapdan I.K., Kargi F., McMullan G. and Marchant R. (2000) 'Biological decolorization of textile dyestuff by *Coriolus versicolor* in a packed column reactor', *Environ. Technol.*, **21**, 231–238.

Laing I. (1991) *Rev. Progr. Colour.*, **21**, 56.

Paar A., Costa S., Tzanov T., Gudelj M., Robra K.-H., Cavaco-Paulo A. and Gübitz G.M. (2001) 'Thermoalkalistable catalases from newly isolated *Bacillus sp.* for the treatment and recycling of textile bleaching effluents,' *J. Biotechnol.*, **89**, 147–154.

Paul R., Thampi J. and Naik S.R. (1997) *Indian Textile J.*, No 2, 21.

Pechenik J. (1991) *Biology of the Invertebrates*, William C Brown.

Razo-Flores E., Luijten M., Donlon B., Lettinga G. and Field J.A. (1997) 'Complete biodegradation of the azo dye azodisalicylate under anaerobic conditions', *Environ. Sci. Technol.*, **31**, 2098–2103.

Reyes P., Pickard M.A. and Vazquez-Duhalt R. (1999) 'Hydroxybenzotriazole increases the range of textile dyes decolourised by immobilised laccase', *Biotechnol. Lett.*, **21**, 875–880.

Robinson T., McMullan G., Marchant R. and Nigam P. (2001) 'Remediation of dyes in textile effluent: a critical review on current treatment technologies with a proposed alternative', *Biores. Technol.*, **77**, 247–255.

Shin K.S. and Kim C.J. (1998) 'Decolorization of artificial dyes by peroxidase from the white-rot fungus, *Pleurotus ostreatus*', *Biotechnol. Lett.*, **20**, 569–572.

Slokar Y.M. and Marechal A.M.L. (1998) 'Methods of decoloration of textile wastewaters', *Dyes Pigments*, **37**, 335–356.

Socha K. (1992) *Internat. Dyer*, **177** (7), 21.

Sumathi S. and Manju B.S. (2000) 'Uptake of reactive textile dyes by *Aspergillus foetidus*,' *Enzyme Microbiol Technol.*, **27**, 347–355.

Walker G.M. and Weatherly L.R. (1999) 'Biological activated carbon treatment of industrial wastewater in stirred tank reactors', *Chem. Eng. J.*, **75**, 201–206.

Yatome C., Yamada S., Ogawa T. and Matsui M. (1993) 'Degradation of crystal violet by *Nocardia corallina*', *Appl. Microbiol. Biotechnol.*, **38**. 565–569.
Zhang F. and Yu J. (2000) 'Decolourisation of Acid Violet 7 with complex pellets of white rot fungus and activated carbon', *Bioprocess Eng* , **23**, 295–301.
Zimmermann T., Kulla H.G. and Leisinger T. (1982) Properties of purified orange II azoreductase, the enzyme initiating azo dye degradation by *pseudomonas KF46. Eur. J. Biochem.,* **129**, 197–203.
Zimmermann T., Gasser F., Kulla H.G. and Leisinger T. (1984) Comparison of two bacterial azoreductases acquired during adaptation to growth on azo dyes. *Arch. Microbiol.*, 138: 37–43.

Index